Build Your Own Car PC

Build Your Own Car PC

Gavin D. J. Harper

McGraw-Hill

New York   Chicago   San Francisco   Lisbon   London   Madrid
Mexico City   Milan   New Delhi   San Juan   Seoul
Singapore   Sydney   Toronto

Cataloging-in-Publication Data is on file with the Library of Congress.

1 2 3 4 5 6 7 8 9 0 DOC/DOC 0 1 0 9 8 7 6

ISBN 0-07-146826-9

The sponsoring editor for this book was Judy Bass and the production supervisor was Pamela A. Pelton. It was set in Minion by Keyword Group Ltd. The art director for the cover was Anthony Landi.

Printed and bound by RR Donnelley.

 This book was printed on recycled, acid-free paper containing a minimum of 50% recycled, de-inked fiber.

McGraw-Hill books are available at special quantity discounts to use as premiums and sales promotions, or for use in corporate training programs. For more information, please write to the Director of Special Sales, McGraw-Hill Professional, Two Penn Plaza, New York, NY 10121-2298. Or contact your local bookstore.

The manufacturers' logos in this book are displayed for the purposes of illustration, and to aid the reader in product identification. No warranty, endorsement or validity is given to the book by any of the manufacturers featured.

The author endorses the products shown in the book, as he has proven them to be a winning combination for a good Car PC setup; however, no responsibility is accepted for changes in product specification or any variations which may result in incompatibility.

The author has used his best endeavors to ensure that the URLs for external websites referred to in this book are correct and active at the time of going to press. However, he has no responsibility for the websites and can make no guarantee that a site will remain live or that the content is or will remain appropriate.

Contents

Contents

Foreword

For the first half of the twentieth century, industry, society and culture were changed forever by the invention of the automobile. The second half of the century experienced a greater shift caused by the global availability of computers. The start of the twenty-first century marks the beginning of an explosion as the two greatest forces of the previous century merge.

New car computing technologies are emerging on a weekly basis. As these technologies surface most people will underestimate the potential by only looking at the obvious features such as navigation, movie playback, mp3 playback, OBD (on board diagnostics), voice recognition and Bluetooth integration. There is so much more on the horizon.

I think most advanced car features will be created by user-designed software packages. Market differentiation for auto manufacturers will not come from who has the greatest factory-installed features, but who complies best with open standards to allow for the global creativity of software developers to enter the car. Instead of replacing the car to get the newest navigation system or car feature the owner will just upgrade the software and processor. For the most part, the computer industry leaders have embraced open standards; however, automotive manufacturers have strongly resisted standards. I hope we can work together to make sure the hood doesn't get welded shut as the car leaves the factory.

I dream of future car computing to be anything the owner or driver desires through open standards. There is no doubt in my mind that if the car computing community can successfully orchestrate a collaboration between current automotive and computing technology leaders this truly will be a twenty-first century explosion. The growth of car computing will increase quality of life, safety, work place efficiency and even help save the environment. Does that sound a little bit crazy? Yes. It sure does sound crazy but it will happen.

Since the advent of computers, car geeks who wanted computer integration have suffered from problems that scared off all but the most determined. These problems have inspired over 500,000 questions and answers on the Mp3Car.com support forums. The forums have attracted millions of geeks from all around the world who still struggle to get a computer in their car. I am so excited to see books like this that simplify the installation process, organize forum data and move car computing into a mainstream hobby.

ROBERT WRAY
Co-Founder and CEO of mp3CAR.com/

Preface

In-car electronics has come a *very* long way. Only a few decades ago "in-car entertainment" was unheard of. Since then fads have come and gone, in-car record players, tape cassettes, video cassettes and eight-tracks are all now unheard of, while radio has endured.

I got interested in Car PCs when I started to look around at "traditional" in-car entertainment. I quickly began to realize that there was little chance of putting together a coherent system to deliver multimedia, navigation and advanced features without boxing myself into a corner by buying from just one manufacturer.

It rapidly became apparent that proprietary standards rule in in-car entertainment and telematics, and that if I wanted an integrated system that fulfilled all my needs – and here's the catch – at a reasonable price, I was going to need to think outside the [bass] box and look at an innovative solution.

The electronics industry is constantly delivering new innovations; new formats come out every couple of years. In-car entertainment has evolved, radio is in a transition period, moving from analog to digital, and when you think about the variety of different media and technologies that can be stored on a 15 cm optical disk, with new technologies such as Blu ray and HD DVD in the pipeline, you begin to realize that a hard-wired, fixed solution cannot compete with a modular, upgradeable, "update the software and it works," flexible solution.

And the beauty is, you can express your individuality through your Car PC setup in the same way as you can with car modifications.

For ages, buying bigger and better, faster and more powerful cars has been the rage; performance has been the order of the day – however, it is dubious whether this can continue indefinitely. Oil prices are soaring through the roof, and despite the reluctance of petrolheads to give up their gas-guzzlers, one day an age may dawn where legislation, prices and physical resources dictate that owning a hydrocarbon-munching monster is no longer cool.

People will always want to be individuals, customizing their vehicle to their own tastes and preferences.

Will car bumper stickers someday read "My other Car PC is a Mac"??

Enter the Car PC – megahertz is the new horsepower, code the new gas.

I will conclude with a short quote from a Forrester Research report:

"Just as the hot rodders of the 1950s begat Detroit's muscle cars of the 1960s, today's telematics hackers portend the future of telematics innovation."

Acknowledgments

A book is never the work of one person; the ideas expressed by the author are always the result of the help and encouragement of a large number of people, an invisible support network, working behind the scenes to make things work.

First of all a major thank-you to Tim Watson from Photomedia UK Ltd., without whose help, the high-quality close-up photography for this book would not have come to pass. If a picture tells a thousand words, then Tim has surely written much more of this book than I.

My fascination with Car PCs started when I was working for Betoddoreven.com. Without the encouragement of Brian Reid, my boss, and the rest of the staff at Betoddoreven.com, this project might never have left the ground. Thanks for all the encouragement you gave me.

I would like to convey my appreciation to all the people who helped me procure components for the build pictured in this book. Thanks to Beth Ellerman at Hitachi. To the folks at Pioneer, Carole Love, and Brendan Sheridan, thanks for all their help. Gratitude to Louise Huang at Travla. Thanks to Gillian Smith and Sam Harmer at Crucial Memory.

Another great big thank-you to Bulent Özen, whose expertise in on-board diagnostics has proved invaluable throughout writing the section on OBD. All of the designs in that section were developed by his firm Özen Elektronik, and I am truly grateful that he has allowed me to reprint his designs.

I would also like to thank Gaynor de Wit, Werner du Plessis, Stuart and Richard Brown, and Fiona Gatt at VIA

My thanks to Robert D. Wray, who has been great to bounce ideas off of and who wrote the Foreword to this book.

Thanks also to Armen and Marina at DigitalWW.

There is bound to be someone who has provided help along the way that has been omitted, my sincerest apologies.

Thanks to Andy Baxter at Keyword and Alan Foster for making the copy edit on this book really painless. Thanks for your patience guys.

Finally, but certainly not least, a big thank you to the folks at McGraw-Hill, to everyone in the office who makes dealing with McGraw-Hill such a joy, especially Anthony Landi and Diana Mattingly.

The biggest thank-you has to go to my amazing editor Judy Bass, who I am totally indebted to for being totally magnificent and making the whole publishing process such a breeze. I am still of the conviction that Judy is one of the nicest people in New York.

Build Your Own Car PC

Chapter 1

Why would I want to build a Car PC?

In the past several years, there has been an explosion in the amount of electronic functions and features that are available to the driver and passengers of the modern automobile.

Satellite navigation is now commonplace with small GPS devices available for only a few hundred dollars while the explosion in in-car multimedia has resulted in some very attractively priced head units coming to market.

Bluetooth phones are now common, allowing you to connect to the Internet through your mobile phone which is still in your jacket pocket.

PDAs allow us to manage important information in the palm of the hand, which we can then synchronize with our home PCs.

With this vast array of products available at a reasonable price, you may well ask what the advantages of building your own Car PC are when a lot of the Car PC's functions can be achieved with individual devices.

What you want, how you want it

When you build your first Car PC, you are creating a unique piece of in-car entertainment that is soooooo flexible, that you can customize it to work in synergy with the way that you drive and use your vehicle. Just as you would customize your PC desktop theme background, screen setting and volume level to suit the way that you work at home or in the office, so you can customize your Car PC to work with you in the way that you drive. Don't put up with what your manufacturer has given you! Original equipment may integrate very nicely with the vehicle but the chances are that its functionality is not cutting edge. It takes a lot of investment for auto manufacturers to develop new products, so when they have something that works, the chances are they are going to cling on to it for a bit. Furthermore, ordering options such as GPS as factory or dealer fit options is usually tremendously expensive. By contrast you can buy PC components off the shelf at attractive prices.

Custom car/Kit car builders

It may be that you have a vehicle like no other, a unique creation that you have spent the last decade building from the ground up. Every last nut and bolt was specified by you, you spent hours searching the Internet for that special custom paint finish and many days trawling breakers' yards for unique components that would make your car like no other. So excuse me for shouting "MUG" when you go and stick an off-the-shelf consumer head unit in it! Why not create an automotive informatics system that is as unique as your creation? By using custom "skins" and graphics for your programs, you can create a look that is in keeping with your vehicle and does it justice.

Customizability

Create something unique that is as personal as you! By designing a system around the way you drive, you can ensure that the features and functionality it contains are right for you. There is no such thing as the average driver; everyone has their own driving styles; different drivers make different types of journeys, and require different types of information to make those journeys run smoothly.

Stay ahead of the curve

Developing an in-car entertainment system is an expensive business for the big boys: vehicle manufacturers invest a lot of money in producing GPS systems, and to see the best return on their profits it is in their interest to sell the product for as long as they can – this does not necessarily meet with the requirements of the power user who wants the latest technology – NOW – in their car.

Integration

At the moment there are so many devices available for in-car use that it is quite possible for the driver to be overwhelmed by the amount of information that is presented. There have been studies about how a driver reacts to different sources of information in a vehicle. Compelling evidence has been presented that a driver's reaction time to events happening on the road decreases as the driver becomes overloaded with more and more information from inside the vehicle. This would seem common sense. By integrating all of these devices into a single console and interface, many of these distractions can be eliminated or at least reduced. As with any technology, it is imperative that the individual uses it in a responsible manner. Watching DVDs while driving is rightly illegal in many localities as well as an example of technology being used irresponsibly. This is not to stop passengers enjoying the in-car entertainment from the comfort of their own seat – as long as it does not distract the driver.

Another pressing case for integration can be seen by using a home multimedia/HiFi/Theatre system as an analog for in-car entertainment and information devices. Some people would prefer a "stack" of separates with each individual device being selected on its credentials for the job. Unfortunately, this not only takes up an enormous amount of space but also paves the way for unsightly cables hanging out from every angle! Fortunately, with a few small compromizes, it should be possible to integrate all of these devices into a single unit.

This is the situation with the Car PC at the moment. It is possible to build a device that will integrate hi-fi functions, navigation, mobile office, Internet on the move and other functions in a single unit with a few small compromizes. Thankfully the concessions made for assimilating all of the devices into a single machine are few and far between and on balance are far outweighed by the benefits.

In each of the sections, we will discuss the advantages and disadvantages of a Car PC-based system over other alternative approaches. Over the next few pages we will discuss some applications that your Car PC will be able to perform.

Audio jukebox

Car PC enthusiasts have been building units to play *.mp3 formats for almost as long as the format has been around. Realizing its popularity, commercial interests responded by releasing *.mp3 capable head units that would read *.mp3 files burnt to CD. Some manufacturers even made hard disk based units. But the problem is that everything is proprietary with these systems. If you want to increase storage capacity you have to buy one of the manufacturers' hard disks in their own caddy, etc. And manufacturers often charge a much greater $/£/¥/€ per mb than were you to buy the hard disk from a computer retailer. By designing your own Car PC system based on industry standard components, you know that it will be easy to replace components in the case of upgrade or failure.

It is certainly the case that *.mp3 files (and other allied formats) have become phenomenally successful over the past couple of years: the meteoric rise of Apple's iPod, in a comparatively short space of time demonstrates that the consumer is ready to take their entire music collection with them wherever they go.

The same can be said for drivers and passengers over time products such as in-car record players, tape players, 8-track, and CDs that have been sold to consumers with varying degrees of success; users have often been forced to compromize on either sound quality or amount of music in order to have the tunes they want, when they want. Listening to the radio is great, but there are certain times when you want to listen to the music *you* want rather than the random choice of some DJ a hundred miles away.

Compromize no more – laptop hard drive technology allows you to store more music than you could ever possibly want to listen to on a small compact drive that is both reliable and rugged in a car environment. Your Car PC interface allows you to access this vast archive of music instantly with ease – cataloging and indexing tracks by song title, artist, album or a variety of other parameters.

Increasingly, users are buying music in digital file formats from the web; the advantages are clear – many files can be stored on a small hard disk reducing much of the bulk of a large CD collection.

So you have an audio jukebox server at home complete with hundreds of thousands of tracks ripped from your own CD collection or bought from the web. Well, why not seamlessly update your car's collection of music via a wireless network, every time you park your car in the garage? By keeping your car's audio collection and home audio collection synchronized you can ensure that any tracks you add to your personal collection are with you wherever you go.

Radio replacement

Conventional audio FM/AM

You might want to retain your existing radio if it is part of a manufacturer's installation; however, if you are sacrificing your radio for your Car PC, do not worry; help is at hand. In this book, we will talk you through the installation of a Griffin Radio Shark AM/FM receiver. This allows you to receive radio on your Car PC. It is simple to install, connects via USB and allows you to receive crystal clear FM radio. The tuner seeks fast, and allows you to store presets along with the name of the station. It also has funky cool blue LEDs in the case – which is nice. Also, imagine that you have to attend a function, but unfortunately, your favorite radio show, or a sports game is being broadcast on the radio. No problem. The Radio Shark allows you to record the radio program on your Car PC, so that you can listen to the show on the way home! Clearly, there are not many in-car radios on the market that allow you to do this!

Digital XM/DAB/satellite digital radio

Depending on where you live, the chances are that you are able to receive some form of digital radio, whether this be XM/DAB or satellite radio. The advantages of digital broadcasting are manyfold – the sound is immensely clear, CD-like sound without crackle or interference. Furthermore, there is a lot more scope with digital broadcasting to send additional information about the track and station. With RDS, used with conventional analog radio, you are limited to basic information such as the station name and genre of music played; however, with digital broadcasting there is the possibility to send diskography information and much more in the form of text and graphics. Once you have built your PC, you will find it very easy to find a compatible digital radio receiver that will interface to your Car PC via USB.

Internet radio

In later sections we will evaluate the options for accessing the Internet from your vehicle. If you are in a wireless Internet (WiFi) enabled area, you can explore the world of Internet radio; this opens up the possibility of listening to radio stations from around the world. To use this feature, you will need to have a stable broadband service, so you may find that you are OK listening to the radio when parked up outside your home or office but not while on the move.

In-car video

Television on the move

Televisions in cars began to be seen throughout the 1990s, but very early installations were CRT based, a little bulky and analog. However, since then technology has improved both in quality and size, and it is now possible to get LCD screens which require a lot less mounting depth. Furthermore, by coupling a Digital Freeview receiver to your Car PC, you can receive crystal clear digital pictures in countries where terrestrial digital television is available. This is yet more "value added" from your Car PC.

DVD

In-car DVD players are a lot more expensive than domestic DVD players, as they are targeted at a niche market. If you are going to buy an "out of the box" in-car DVD player, the chances are you will be paying through-the-nose. However, by contrast, DVD drives for computers are relatively cheap; you may even have one lying around at home. By adding a DVD drive to your Car PC and some decoding software, it is possible to play DVDs from all regions in your car. Furthermore, you can use your Car PC to create backups of your DVD disks onto your Car PC hard disk, allowing your passengers to watch DVDs on the move without your having to carry bulky disks in the car. This of course is subject to your owning the copyright on the DVD. Cloning DVDs from the video star for use on your Car PC is strictly illegal.

Divx

Divx is a great file format that allows large, bulky video files to be compressed into a relatively compact file size. The advantage of this is that with the correct software installed on your Car PC, you are able to store oodles of video on your Car PC hard disk. Whether this is for kids' films or action movies, the long journeys will certainly seem a lot shorter for your passengers. Again, this is a feature that is hard to find in conventional in-car audio, but – as we have seen with the introduction of in-car mp3 head units – where Car PCs lead, commercial manufacturers will shortly follow.

Mobile office for the portable professional

Wouldn't it be handy if you could keep in contact with your world while on the move. By installing a Car PC it becomes easy to "take your office with you," allowing you to answer your emails, send and receive faxes, compile and edit documents, all from the comfort of your car. The thing is, you don't always want to carry a laptop with you wherever you go, and laptop computers are a target for thieves; it is a lot harder to steal a PC that is integrated into your vehicle without taking the whole vehicle.

Integrating with existing in-car radio equipment

You might have just purchased an expensive high-end motor, only to find that the radio is integrated into your car. Removing it would leave ugly scars along with steering wheel controls and other accessories which are nice to have but next to impossible to remove. Your head unit hasn't got any auxiliary inputs and you simply don't want to hack your car to bits.

No problem! In this book I am going to show you how to install the Griffin RocketFM into your Car PC, which allows you to "transmit radio" to your in-car setup. This way you just set up your Car PC audio as a preset on your radio, and you select your Car PC audio much as you would your favorite radio station. Because there are no wires connecting to your audio system, you do not need to break into your wiring loom, and your manufacturer's warranty remains intact.

The video system can be run separately and unobtrusively with the only connection to the car's wiring loom being that to a 12 V feed, which can be done with minimal disruption, *or* if you really don't want to touch your car's wiring loom, you can always power your install from a cigarette lighter socket.

So – again – why should I build a Car PC?

Oh yeah . . . and if I haven't already mentioned it, it's also a lot of fun!

Chapter 2

Buying your components

Let's hit the shops folks and shop till you drop. By indulging in a little impromptu retail therapy you could have all the components you need to build an ultimate Car-PC system by the end of this chapter. Fortunately, with the exciting retail opportunities brought by the Internet all of this can be done from the comfort of your office chair. So take a pew folks, grab your wallet (or your parents' while they aren't looking) and let's start surfing.

What follows is my "Ultimate Car PC List." It is possible to use other components, but what I am presenting below is a proven solution that is guaranteed to work, and it presents what I believe is the best that the market has to offer. With each selection, I have highlighted the features and benefits and why I believe that the selection I have made is best for the job.

Don't get me wrong; there are other solutions to installing a PC in your car; certainly some of the cases that are not constrained by the "1 DIN" size limitations can offer room for much faster motherboards, etc. But my personal feelings, as I have mentioned earlier, are that for a Car PC to warrant the title "Car PC" it has to be able to integrate with the car.

You probably have some techno-junk lying around from various projects. Think of innovative ways of integrating this into your Car PC setup; old laptops can be cannibalized for their drives, USB peripherals can often find another home in the car – the list is endless.

Travla/Casetronic C1xx Case

 www.travla.com

7

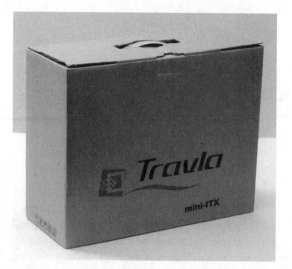

Figure 2.1
Travla C1xx Case box shot

Features and benefits

- Low noise and lightweight
- Compact – same size as a car stereo
- Accepts 0.8" height SDRAM memory
- Accepts slim type 2.5" HDD
- Accepts slim type media drive
- Excellent heat dissipation through extensive sets of vents
- Convenient carrying handle
- Available in black or silver color
- Material: Lightweight aluminum alloy and CRS
- Finish: Aluminum front panel with anodized finish
- Dimension (W × H × D): 7" × 2" × 10" or 177.80 mm × 50.80 mm × 254.00 mm
- Cooling: 40 mm × 40 mm × 10 mm
- Main Board: VIA mini-ITX EPIA motherboard (low profile heat sink required)
- Drive Bays: 1 slim media drive and 1 slim Hard Drive (2.5")
- Internal Power Supply: Built-in 60W DC power board
- External AC Adaptor: – Input (AC 100~240 V) – Output (DC 12 V@5 A) 60 W

What's in the box?

- Travla aluminum case
- Bezels to suit variety of motherboards
- Internal PSU
- External PSU
- Standard IDE Lead
- Modified IDE Lead with termination for laptop connectors
- Array of screws and fixings

Cane CB4 or C156 SP13000

The Travla case really warrants the title of "King Car PC Case." Unlike other "pretenders to the throne," the Travla case actually fits in the same space as a "single DIN" car stereo, which means that it will fit the vast majority of vehicles.

Incidentally, you will find the Travla case is longer than most car stereos, but in most cases a space can be cleared relatively easily behind the dashboard. If you really, really cannot make the room, then you still have plenty of other mounting options left. You can find DIN brackets that will allow you to sling the Car PC underneath your dashboard.

If your car stereo is fitted in a custom space (as some manufacturers do not adhere to standard sizes) you will probably find that you can procure an adaptor plate from your local car spares shop. This fits over the large hole in your dashboard, and provides you with one or two standard size DIN slots. The surrounding fascia is designed to blend in with your dashboard, giving an "original equipment" feel to the installation.

One of the things you need to bear in mind is that even though the Travla case is designed to work with minimal ventilation, it does still need a little room to breathe, so try to clear the way behind your car stereo slot in order to allow fresh air to cool the motherboard.

VIA EPIA SP 1300 Motherboard

 www.via.com.tw

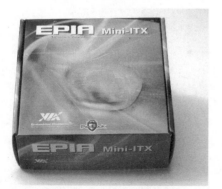

◀ Figure 2.2
*VIA EPIA SP 1300 Motherboard box shot
(Tim Watson, Photomedia UK)*

Features and benefits

- Chipset & Core Functionality

  - O VIA C3 1.3 Ghz Processor
  - O CPU Voltage Monitoring
  - O VIA CN400 North Bridge
  - O VIA VT8237 South Bridge
  - O 1× DDR Memory Slot 266/333/400 Mhz (up to 1GB)
  - O Integrated VIA UniChrome™ Pro AGP Graphics
  - O UniChrome MPEG-IV Accelerator
  - O VIA VT6103 10/100 Base-T Ethernet
  - O VIA VT1617A Integrated 6 Channel Audio (AC'97)
  - O VIA VT1623 TV Encoder
  - O VIA VT6307S IEEE 1394 Firewire Encoder
  - O Award BIOS
  - O 4/8 Mbit flash memory
  - O Wake On Lan, Keyboard Power On, Timer Power On
  - O System Power Management
  - O AC Power Failure Recovery

- Internal Input/Output Ports

  - O 1× PCI Slot
  - O 2× Parallel ATA IDE Slots (UltraDMA 133/100)
  - O 1× CIR connector (Keyboard/Mouse)
  - O 1× DB25 Parallel Port Connector
  - O 1× Wake On LAN Connector
  - O 2× Fan Connectors (CPU Fan/System Fan)
  - O 1× I_2C Connector
  - O 1× LVDS/TTL Module Connector (Optional Extra)
  - O 1× +12V Power Connector
  - O 1× ATX Power Connector

- External Input/Output Ports

  - O 2× Serial ATA Connectors
  - O 1× PS/2 Mouse Port
  - O 1× PS/2 Keyboard Ports
  - O 1× RJ 45 Lan Port
  - O 1× DB9 Serial Port
  - O 2× USB v.2.0 Ports
  - O 1× VGA Port
  - O 1× PCA Port (can either be used for Composite TV-OUT/SPDIF)
  - O 1× S-Video Port
  - O 3× Audio Jacks with Smart 5.1 Support

- ☐ Line Out (or Front Left/Front Right)
- ☐ Line In (or Surround Left/Surround Right)
- ☐ Mic In (or Center/Sub)
- ☐ Physical Characteristics

- 17cm × 17cm

  - O Operating Temperature 0 degrees C – 50 degrees C
  - O Operating Humidity 0% – 93% relative humidity non-condensing
  - O 6 layer Printed Circuit Board

What's in the box?

- VIA Mini ITX Motherboard
- Quick Installation Guide
- ATA 133/100 Parallel ATA IDE Cable
- Driver and Motherboard Utilities CD
- Input/Output Bracket

The problem with cramming so much technology into such a tiny space is the fact that things start to get really hot!!! Sticking a P4 in a case not much bigger than a car stereo is going to result in total meltdown!!! In order to build a PC that is robust, and is not going to wimp out at the faintest hint of a hot spell, we are going to use a chip that integrates VIA's "Coolstream" architecture.

This uses a 90 nm Silicon on Insulator process to create fast processors that run cool. The benchmark we are looking at here is nothing as unrefined as raw power, but instead "performance per watt." In this respect, VIA lead the market, providing processors that produce an acceptable speed, while drawing very little power.

The VIA CN400 Northbridge is a very competent chipset dealing with all of the "fast" interface operations such as video, TV-OUT and memory interfacing.

The VT8237 Southbridge Chipset deals with all of the slightly slower interfacing functions such as sound capabilities and disk interfacing.

We will discuss the North and Southbridge Chipsets in much greater detail in Chapter 3.

The Unichrome Pro is a very capable graphics engine which deals with both 2D and 3D display. The chipset is TV-OUT capable and ready to deal with 1080p High-Definition TV. One of the real bonuses for in-car use is the fact that the chipset decodes MPEG II video directly, taking strain off the processor. Another nice feature of the chipset is hardware de-blocking of video, which on lower quality video clips, will reduce the "blocky" pixellated appearance.

Pioneer DVR-K04 DVD media drive

◀ Figure 2.3
Pioneer product shot (courtesy Pioneer)

 www.pioneer.com

Features and benefits

The drive supports the following formats:

- CD-Audio
- CD-ROM
- CD-R
- CD-RW
- DVD±R
- DVD+R DL
- DVD±RW

The Pioneer DVR K04 comes with a 2Mb buffer to ensure error-free disk writing.

The drive reads at the following speeds:

- DVD-ROM 8X
- CD-ROM 24X

The drive writes at the following speeds:

- DVD±R 8X
- DVD+R
- DVD+R DL 2.4X
- DVD±RW 4X
- CD-R 24X
- CD-RW 24X

What's in the box?

- Bare Pioneer Laptop Drive

There are so many different conflicting formats for DVD writable drives that it is easily possible to become confused. Thankfully, the Pioneer DVR K04 comes with support for all the major disk formats (with the exception of DVD RAM) and does it all in a stylish compact package. One of the major selling points is the fact that it is slot-loading, meaning that your Car PC will take on a high-end quality appearance. You wouldn't expect to have a disk tray on a top-notch piece of ICE, so why would you want a disk tray in your Car PC? Loading a slot-loading drive is *sooo* much easier, and has infinitely more cool than fiddling to get a DVD on a spindle on a flimsy plastic tray. Furthermore, trays can easily come off if abused even the slightest bit. With a slot-loading drive you are safe in the knowledge that you have no disk tray to break off!

Writing at a blindingly fast 8x on DVD±Rs this drive definitely has my vote!

Hitachi Endurastar 30GB hard disk

 www.hitachi.com

Figure 2.4
Hitachi Endurastar HDD
in antistatic bag

While this may not be the largest or fastest laptop hard disk on the market, it is a wise purchase as it has been designed specifically to work in harsh environments where a rugged mechanism is required. In layman's terms – it's solid. This drive will take a lot of abuse compared to your standard drives which will wimp out at the first sign of a bit of vibration or while working in "sub-optimal" climates. Some of the technologies employed by Hitachi to produce such a rugged drive include:

- "Fly Height Compensation": this adjusts the height that the head flies above the platter to accommodate changes in external air temperature and air density and pressure caused by changes in altitude.
- "Built-in Shock Sensor": The Endurastar range incorporates a built-in shock sensor that parks the heads preventing a collision with the platter at the first sign of trouble.
- "Enhanced Moisture Control System."

Features and benefits

- 30 GB Maximum Capacity
  - O Equivalent to:
    - □ 7500 4 minute *.mp3 songs
    - □ 2.7 hours of DV25 video
    - □ 4 hours MPEG-II video
    - □ 15 interactive games
- Fluid Dynamic Bearings for quiet operation
- −20 degrees C to + 85 degrees C operating temperatures
- Load/Unload Technology – reduces disk wear and maintains data integrity
- Built in shock sensor – for performance in rugged environments
- Fly Height Compensation (to adjust to differing air temperatures and densities)
- 2.5" Laptop form factor
- Enhanced moisture control system to prevent corrosion in extreme environments
- Non-operating shock tolerance 800G/1ms
- Operating shock tolerance 250G/2ms
- Designed for automotive applications

What's in the box?

- Hitachi Endurastar Drive in Antistatic Wrap

Griffin Radio Shark

If you are ripping out your old FM radio to replace it with a Car PC, do not worry, you will not lose radio functionality. I am going to show you how to tune in to AM and FM radio using the Griffin Radio Shark receiver.

Features and benefits

RADIO FEATURES

Tuner Frequency Range –

- FM: 87.5–107.9 MHz
- AM: 530–1710 kHz
- Built-in AM/FM Antenna
- Tuning Display – Software based

RECORDING FEATURES

- Record Level Control – Automatic
- Up to 44.1k Recording (CD Quality)
- Three presets and custom settings available

▲ Figure 2.5 (i)
Griffin Radio Shark front

▲ Figure 2.5 (ii)
Griffin Radio Shark rear

GRIFFIN TECHNOLOGY

What's in the box?

- Griffin Radio Shark Hardware
- USB Extension Lead
- Driver CD
- Manual

Griffin RocketFM

Don't want to hack into your car's wiring loom but still want a Car PC? Worried about invalidating your car's warranty? Not confident with auto electrics? Don't worry, you don't have to be. I am going to show you how to install the Griffin RocketFM, allowing you to transmit all of your Car PC's audio signals to your standard car stereo without touching the wiring loom.

By powering your car stereo from a cigarette lighter, you can achieve Car PC functionality, but without touching your vehicle electrics.

For owners of new vehicles with a warranty they do not want to invalidate, this is a real boon.

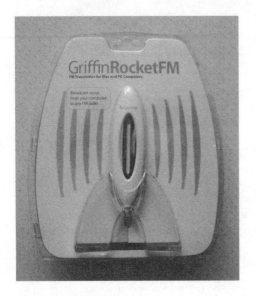

▲ Figure 2.6 (i)
Griffin RocketFM II

▲ Figure 2.6 (ii)
Griffin Rocket FM II rear

GRIFFIN TECHNOLOGY

Features and benefits

Technical Specifications

● Compact size: 4.2" (107.5 mm) × 1.2" (31.6 mm)
● Built-in antenna
● Power requirement: powered USB port, 100 mA (low power)
● Modulation: FM Stereo
● Frequencies: 88.1–107.9 MHz
● High stability crystal oscillator, phase-lock loop control

- Stereo separation: greater than 45 dB
- Signal distortion: less than 5%
- Frequency response: 50 Hz to 15 kHz
- Operating range: 10–30 feet, depending on the quality of the FM radio receiver (limited by FCC regulations)
- FCC compliant
- Cable length: 3 feet (0.9 m)

WARNING !!!

FCC COMPLIANCE

This device complies with Part 15 of the FCC Rules. Operation is subject to the following two conditions: (1) this device may not cause harmful interference, and (2) this device must accept any interference received, including interference that may cause undesired operation.

What's in the box?

- Rocket FM Hardware
- USB Extension Lead
- Driver CD
- Manuals

Griffin Air Click USB

The Griffin Air Click USB allows you to control your Car PC from a key fob sized remote control. This can be strapped to your steering wheel or placed conveniently for fingertip control. The remote comes with five buttons that have most of the basic controls you would want for audio playback. Because it is based on RF technology, not infra-red, you don't even need to point the thing at your Car PC for it to work.

Features and benefits

- Operating range: Up to 60 ft.
- Transmission method: Radio Frequency
- Frequency: 433.92 MHz
- Remote battery: CR2032 3 V
- Remote dimensions: 2.75" × 1.25" × 0.5"

(not including belt clip; 0.75" max depth including belt clip)
Remote weight (including battery): 0.8 oz

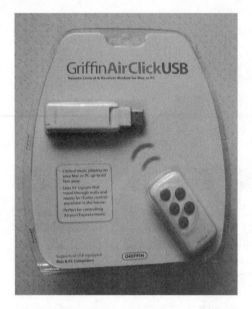

▲ Figure 2.7(i)
Griffin Air Click USB II

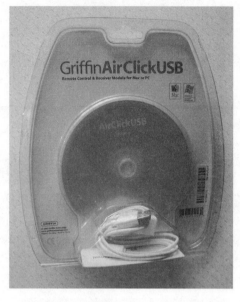

▲ Figure 2.7(ii)
Griffin Air Click USB II Rear

GRIFFIN TECHNOLOGY

What's in the box?

- Griffin Air Click USB Hardware
- Griffin Air Click Remote Control
- Drivers CD
- Instructions

Griffin Total Remote

Ever seen the James Bond film where the super smooth secret agent whips his PDA out of his pocket and steers his car in the car park to enable him to make his getaway?

Well, I'm not going to show you how to do that, and besides, the feds might get annoyed if you ran someone over, but what *I will* show you, is how you can control the functions of your Car PC through a fully skinnable interface on your PDA.

◀ Figure 2.8
Total Remote product shot

Features and benefits

- Allows you to control infra-red ready devices from your PDA (not just your Car PC)
- Extends the range of your PDA's infra-red system
- Does *not* use IrDa, but uses proprietary system for greater remote accuracy

What's in the box?

- Total remote driver CD
- Total remote infra-red booster hardware
- Instructions

Özen Elektronik mOByDic OBD Interface

Figure 2.9
Özen Elektronik mOByDic interface (courtesy Özen Elektronik)

The Özen Elektronik mOByDic interface allows you to tap into your car's on-board diagnostic system, and find out a great deal of information about how your car is performing and running.

The hardware is available as either a circuit you build yourself, bare PCB, or "ready built" solution.

While this is not as straightforward as some "out of the box" solutions, you do have a great degree of flexibility and compatibility with many OBD interfaces. You have the option of building it yourself and saving money, or buying an off-the-shelf solution.

Also, because the interface is not proprietary, it is well supported by a wide range of software.

Features and benefits

- Allows you to diagnose problems with your car
- Ideal tool for power-tuners
- Enhance your car's performance
- Monitor your car's "system variables"

Supports the following OBD interfaces:

- ISO9141-2
- KWP2000
- J1850-PWM
- J1850-VPWM
- CAN-BUS

Crucial memory 1GB compact flash card

As an alternative to a hard disk in your Car PC, we are going to discuss how you can boot from solid state memory, giving the advantages of having no moving parts and enhancing reliability significantly in mission critical applications.

Figure 2.10
Crucial memory 1GB compact flash card (Tim Watson, Photomedia UK)

Features and benefits

● No moving parts
● 1GB of solid state storage

Andrea Electronics DA-350 Auto Array microphone

Voice recognition in-car is notoriously hard to achieve with a PC and standard microphone. The problem is not so much with the speech recognition software, but with the microphone and its ability to deal with ambient noise.

By replacing a standard microphone with a linear array microphone with digital signal processing, we can cut a lot of noise out of the equation, making the speech recognition software you are using *much* more accurate.

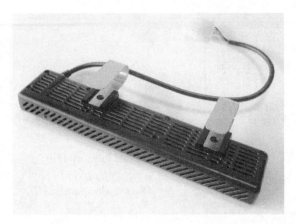

Figure 2.11
DA350 Auto Array microphone

DA-350 HANDS FREE
LINEAR ARRAY MICROPHONE

Features and benefits

- Sensitivity: (0 dB = 1 V/Pa; f = 1 kHz) 63 mV 4 dBV ± 2 dB
- Supply Voltage: 12 V car battery low current consumption
- Directivity: Super Directivity (DSDA)
- Output Line Level: 2 V RMS Max.
- Output Impedance: <100 Ohm
- Equivalent Noise Floor: Max. 30 dB SPL
- Max. Input Sound Level: 110 dB SPL (1 kHz, THD <2%)
- Max. Storage Ambient Temperature Range: −40 to +85°C
- Max. Operation Ambient Temperature Range: −25 to +70°C

What's in the box?

- Andrea Electronics Auto Array
  Microphone with Headvisor Clips
- USB Audio Interface (Optional)

Chapter 3

Building your Car PC base unit

If you are familiar with building computers, building your Car PC is no different than building any other PC except that everything is that much smaller and there is less space for everything to fit! This is going to be a real test of your manual dexterity!

If you are unfamiliar with computers, then do not worry. Everything in this guide is covered in a step-by-step manner with meticulous detail. You need nothing more than a copy of this book, the components listed and a few everyday tools to assemble your working Car PC.

It also doesn't matter if your shed doesn't resemble Home Depot, you don't need a tremendous amount of kit to assemble a Car PC.

Taking antistatic precautions is a foregone conclusion when handling any sensitive electronic equipment. You can use a wrist strap; however, regularly touching a grounded object to dissipate any static electricity and being careful when handling boards should also suffice.

In terms of tools that you will need for assembly, I found that a basic set of the following was sufficient to put together my Car PC. These are shown in Figure 3.1.

In the toolbox

- Phillips head small jeweler's screwdriver
- Short stubby Phillips screwdriver
- Long nose pliers

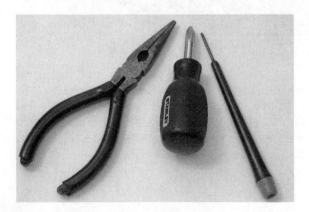

Figure 3.1
Tools of the trade (Tim Watson, Photomedia UK)

Gaining access to the Car PC case is thankfully very easy – Travla have provided two thumb-screws for the rear which make life a lot easier if you are going to make regular adjustments to your Car PC. The first step of our operation is to remove these two screws (see Figure 3.2).

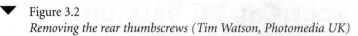

▼ Figure 3.2
Removing the rear thumbscrews (Tim Watson, Photomedia UK)

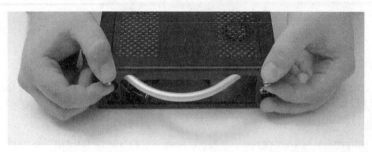

Now put the screws somewhere safe. Even though these are fairly large, we will be dealing with screws much smaller, so maybe now is the time to fetch yourself a small bowl for all the miscellaneous hardware. Now that the rear thumbscrews have been removed we can pull the case apart quite easily. Travla have even provided a natty little silver handle to make the job fool-proof. Just support the external casing at the front and pull the inside out!

▼ Figure 3.3
Pulling the case apart (Tim Watson, Photomedia UK)

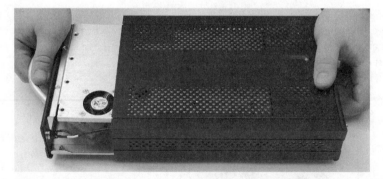

Our first job of the day is to remove the bezel at the front that shrouds the "Media Drive slot." This slot is the size of a standard laptop media drive: the laptop drives are used because they are low power consumption and a lot slimmer than desktop $5\frac{1}{4}$" drive bays. The fascia blanks-off the slot in the front of the case – removing it allows us to install our DVD Media Drive [see Figure 3.4(i) to (iii)]. This is a simple case of removing two screws on what we will refer to

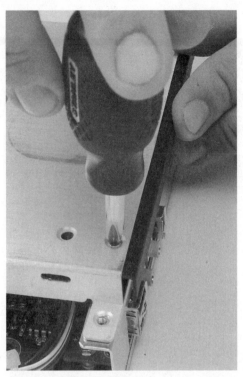

Figure 3.4(i) to (iii)
Removing the media drive bezel
(Tim Watson, Photomedia UK)

(i)

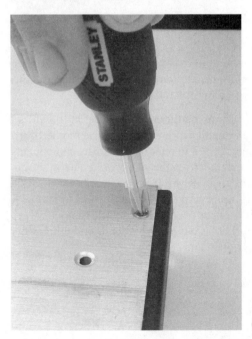

(ii)

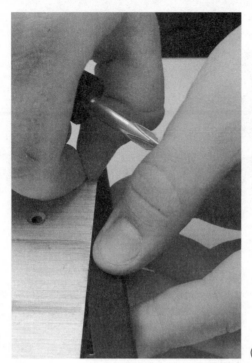

Figure 3.4 (Continued)

(iii)

as the "Media Tray." This is the top piece of aluminum to which we will mount our media drive and hard disk or solid state storage: this is separate from the main case chassis. Once the two screws have been removed, pull the bezel away and retain it for later use.

Now we are going to need to get familiar with our motherboard. Figure 3.5 provides a pictorial representation and Figure 3.6 a diagrammatic representation of the motherboard. You might want to bookmark this page as you will need to refer to it to locate components. The motherboard in the picture differs ever so slightly from the version you have in the box in that its heatsink has been removed for clarity to allow you to get a clearer view of all the components. Take YOUR motherboard, and compare it to the picture. Spend a little time to familiarize yourself with it and identify all of the major components. A little time spent "getting to know" your motherboard now is worthwhile, as it is much harder to find things when it is crammed into the tiny case with cables left, right and center.

▼ Figure 3.5
View of VIA EPIA SP Mini Itx motherboard (courtesy VIA)

▼ Figure 3.6
Diagram of VIA EPIA SP Mini Itx motherboard (courtesy VIA)

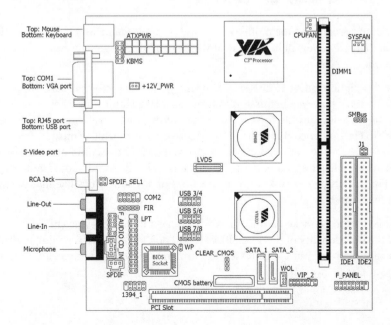

One of the first things that strikes you when you look at the motherboard is how many features are crammed into such a small space. There are many standard ATX motherboards available for desktop PCs that carry a lot less features in a much larger space.

One of the areas where the Mini Itx motherboards from VIA really excel is in the connectivity department. As a result of tremendous integration and EVERYTHING being on-board, the VIA EPIA series always comes with a plethora of rear panel connectors. These will be the things that you use to connect your Car PC to the "outside world" and all your devices in-car. Take a look at Figure 3.7 to explore some of the connectivity options of your Car PC.

▼ Figure 3.7
Rear panel connections (courtesy VIA)

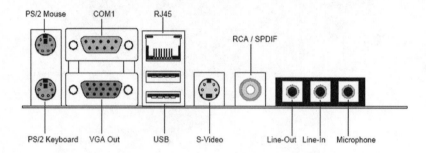

Motherboard walk around

Let's work our way around the motherboard and pick out the major components. If you are looking at Figure 3.5 we will be starting at the top and working our way around in a clockwise fashion.

The first description refers to the single memory slot. The motherboard supports up to DDR400 memory, however, in this application you may be limited by what memory variants are available in a low profile format. In simple terms, low profile memory means "short memory." To clear other components in the Travla case, our memory must be 0.8" or shorter (20mm if you are using metric); because "short memory" is not common, you may struggle to buy it in the higher speed variants. A small sacrifice in speed should not matter here though. Forget putting heatsinks on your memory or anything fancy like that because there just isn't the room here!

The next feature to be picked out is the Northbridge and Southbridge chips. These provide much of the functionality of the VIA EPIA SP. You don't need to do anything to these, other than install the drivers, but I shall review their operation to help give you an understanding of how your Car PC works.

To explain what the Northbridge and Southbridge do, we need to know a little bit about processor and chipset architecture. If you take a look at Figure 3.8 you will get a better understanding of the interactions between the processor and the chipset.

▼ Figure 3.8
Diagram of VIA EPIA SP chipset (courtesy VIA)

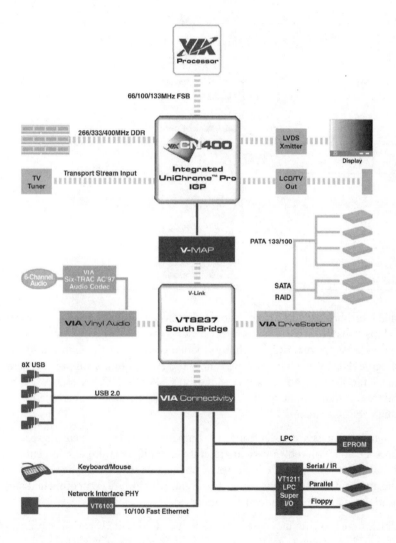

The Northbridge chip deals with all of the super-high speed hardware such as graphics and memory interfacing, while the Southbridge chip deals with applications that require a little less bandwidth.

We can see from the diagram that the CN400 chip is looking after interfacing with the memory, LVDS output, VGA out and TV Tuner. These operations all demand a lot of the processor's power as they are transferring large amounts of information.

The CN400 Northbridge chip provides an interface to the Southbridge chip which requires far less bandwidth to carry out its operations.

The CN400 Northbridge chip is responsible for stunning onboard graphics in a small package. The chipset shares memory with the processor – common practice for on-board graphics, and provides support for high definition video courtesy of the Chromotion CE Video Display Engine. HDTV is taking off big time – the chipset offers support for new HDTV ready digital file formats, for example Microsoft's® WMV HD. As HDTV PC interfaces become more common on the market, it should be possible to watch high definition television through your Car PC.

The integrated graphics are never going to compete with the latest cutting-edge graphics cards. However, you would be a very ungrateful person if you were to complain, when you consider that this whole package is delivered on a 17×17cm motherboard with ultra low power consumption and heat generation. The graphics really are *very* competent and for Car PC applications are more than adequately specified.

The S3 Unichrome Pro Graphics core integrates separate 128bit 2D and 3D graphics engines, and the thing that makes it great for our Car PC application is that decoding of video files such as MPEG 2 & 4 is done by the graphics engine rather than the processor, freeing the processor to do other important tasks! You might like to think of it in the analogy of modern business practice as "outsourcing". This means that the processor can make better use of those precious processor cycles.

UniChrome Pro

The Southbridge deals with all of those tasks that require a little less bandwidth. The VT8237 gives the motherboard support for Parallel ATA hard drives, and unusually for such a small board, Serial ATA and RAID support. The Southbridge also provides 6 track AC'97 audio courtesy of VIA's Vinyl Audio chip standard.

Again, the AC'97 audio is not going to compete with the latest and the greatest, but the quality far exceeds that of your "average" car stereo. If you really have a bank balance as big as the *Titanic* you could splash out on the functionality offered by a 7.1 sound card such as the M-Adio Sonica Theatre, or the THX certified SoundBlaster Audigy 2 Notebook.

In this application, we will not be able to make use of the RAID functionality as there is not room in the case for more than one hard drive.

Nestling next to the BIOS backup battery is the SATA. The Serial ATA is an option for interfacing with hard disks. However, the availability of laptop Serial ATA hard disks is low at the moment, and our Hitachi Endurastar hard disk was selected on the grounds that it is very durable and robust and ideal for Car PC applications. If you wanted an alternative drive, you could consider the Hitachi Travelstar ranges, which are still *very* durable, and available in sizes up to 100Gb; however, they still do not quite offer the same level of shock protection as the Endurastar.

Next we move on to the Parallel ATA IDE ports. We will be using both of these, one for our hard disk and one for a media drive. These will be blue on your motherboard. The connectors are "keyed" which means that you can only insert the cables in one way. Take a little time to look at the connector of the motherboard, and you will see that there is a "slot" that is missing from the rectangular outer envelope of the connector. This slot mates with a plastic protrusion on the IDE cable. Alternatively, there is one of the centrally located pins which is removed. By "plugging" up one of the holes on the IDE cable, manufacturers can ensure additional protection against incorrect insertion. When you are building your Car PC, you will need to remember this and ensure the cables are inserted correctly.

Moving around another step we see the headers for the 3x USB connectors (which gives us a total of six additional USB ports). We will use one of these connectors for the front panel USB of our Car PC, the others are redundant. Saying that though, with a little ingenuity there is nothing to stop you extending these headers with ribbon cables outside of the case as Car PCs with many accessories tend to be *very* USB port intensive.

The next thing that we encounter is the PCI slot. It is probably best to forget about this and not have any designs on it, as within the Travla case there is simply no room for PCI cards; even with the plethora of right-angled adaptors and the like there is no way anything else can be squeezed in the case.

The next label refers to the 1394 header on board which we use for the front panel Firewire connector. If you are going to use an external hard disk you can use Firewire to transfer files at speed between your Car PC and drive interface.

Now we come to the rear panel connectors. The audio jacks are fairly self-explanatory; we can use them to connect to an amplifier or auxiliary input of a car head unit.

Next up is the RCA jack. This can be configured either to output digital audio or to output composite video. In a car that already has existing *video* monitors, you might like to use this jack to provide a feed for existing composite monitors.

The S video port allows you to take a high-quality video feed for *video quality* monitors.

Next up is the LAN connector. You can use this to connect your Car PC to your home network for transfer of files and synchronization: however, a much sounder idea would be to employ a wireless network which will prevent the need for trailing cables.

The two USB ports will be in demand for the plethora of accessories for your Car PC.

The VGA port we will use to connect to an LCD monitor. This will be of a much *higher quality* than the *video quality* monitors used in manufacturers original equipment in car installations.

Although the COM port may seem old and outdated it is actually very useful for a Car PC setup as it allows us to interface with older GPS devices and OBD-11 interfaces.

The PS/2 keyboard and mouse ports require little explanation. We will use these initially for testing and debugging purposes: however, you will find these largely redundant if you intend to use a touchscreen.

The VIA C3 Processor has been designed to power small form factor ×86 architecture devices. It employs VIA's Coolstream technology to ensure that the user gets the best performance per watt with minimal heat produced. This is where the VIA EPIA boards *really* win. They will not drain the juice from your battery and will not produce excessive heat causing them to fail prematurely.

For processor junkies some notable features include:

- Sixteen pipeline stages
- SSE multimedia instructions
- StepAhead™ Advanced Branch Prediction
- Efficiency-enhanced 64KB Full-Speed Exclusive Level 2 cache with 16-way associativity
- A full-speed Floating Point Unit

Mounting the motherboard

Now we have familiarized ourselves with the motherboard and discussed its features and functions, we shall move swiftly onwards to installing it in the case.

The first important step we need to take before mounting the motherboard is to replace the heatsink. You should be supplied with a replacement heatsink in the Travla hardware package that comes with the case.

The reason for replacement is clear to see when we hold the two heatsinks close to each other.

WARNING !!!

Be aware that removing your motherboard's original heatsink will invalidate the warranty.

▼ Figure 3.9(i)
Comparison of heatsinks (Tim Watson, Photomedia UK)

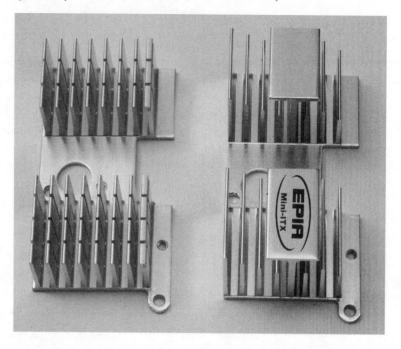

▼ Figure 3.9(ii)
Comparison of heatsinks (Tim Watson, Photomedia UK)

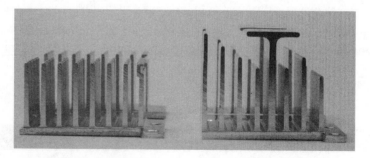

The standard heat sink is *much* bigger than the replacement. The Travla case has an additional fan in the media tray to keep air circulating and make things cool, which compensates for the fact that the replacement is a little smaller. Note, however, that the manufacturer will not honor warranty claims where the heatsink is replaced, and there is always the possibility of damaging the chips (both the CPU and Northbridge), so at this point make sure you are confident with what you are doing.

There is little to go wrong, but the mistakes you make *could* be catastrophic.

Take the motherboard out of the box. It will look something like this (see Figure 3.10)

You need to look at the large aluminum heatsink with the VIA legend on it. Located are the two plastic spring plugs.

Figure 3.10
Motherboard with old heatsink installed (Tim Watson, Photomedia UK)

Now flip the board over and identify the plastic lugs on the bottom of the board. They should be fairly obvious; referring to Figure 3.11 should aid you in identifying the lugs.

The two plugs protrude out the other side. To remove them, you will need to squeeze the splayed plastic lugs together carefully with a pair of long-nose pliers, and give them a little push until they emerge from the other side. Be *incredibly* careful with the long-nose pliers, so that you do not scratch the circuit board. Furthermore, the litte lugs go flying off at high velocity. You can do one of two things. You can either wear goggles, *or* hold the lugs on the other side of the board. Retain the clips.

Figure 3.11
*Underside of
motherboard showing
locations of plastic
lugs (Tim Watson,
Photomedia UK)*

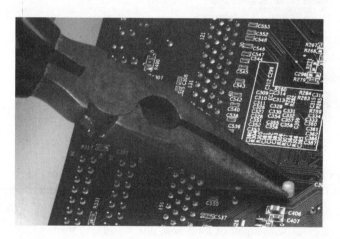

Figure 3.12(i)
*Removing lug 1 (Tim
Watson, Photomedia
UK)*

Figure 3.12(ii)
*Removing lug 2 (Tim
Watson, Photomedia
UK)*

WARNING !!!

Do this next bit REALLY *really* carefully as there is the potential to screw a lotta things up and your motherboard could end up a vegetable!

You will now need to carefully prise the old heatsink from the board. You may need a small implement such as a screwdriver to give you a little leverage. Whatever you do, don't scratch the PCB or start knocking off any of the little itty bitty surface mount devices – without them your board is going to malfunction.

Another thing to be wary of is the capacitor in the middle of the heatsink. When removing the board, be very careful not to apply a lateral force to the heatsink which could damage or otherwise remove the capacitor – you don't want to be paying an impromptu visit to the trash can.

Once the old heatsink has been removed, remove all of the heat-conductive goo from the microchip, leaving a nice clean surface. The new heatsink comes with two pads. You need to peel the backing tape from both sides.

You now need to carefully align the new heatsink with the holes from the old heatsink. The capacitor in the middle of the two chips is a great aid to locating the heatsink. Remember to try and keep a little separation between the chips and the heatsink until you are absolutely sure you are in the right place.

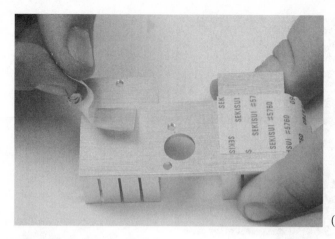

Figure 3.13(i) and (ii)
*Removing tabs from
heatsink (Tim Watson,
Photomedia UK)*

(i)

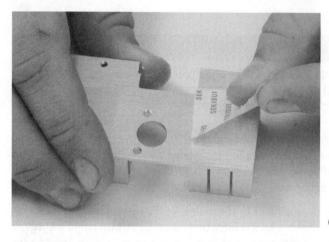

(ii)

WARNING !!!

Do not repeatedly stick and unstick the heatsink as you will degrade the mating surfaces treated with heat transfer compound and cause the thermal joint to be poor and unreliable.

Now look at Figure 3.14. This is what your motherboard should look like with its new heatsink in place.

Now that the motherboard is prepared we can put it safely to one side, as we are going to get the rest of the case prepared to accept it.

Figure 3.14
*Mini Itx with new
heatsink installed (Tim
Watson, Photomedia
UK)*

The next job of the day is to remove the "Media Drive Tray." Looking at the piece of aluminum from which you removed the media drive bezel, you will see it is attached to the main chassis by four screws. These screws are located to the front of the case, either side of the media drive, and to the rear of the case near the rectangular hole through which the motherboard's ports will protrude.

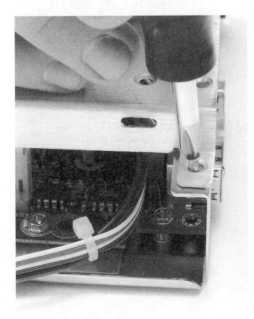

Figure 3.15(i)
*Removing the first media tray screw
(Tim Watson, Photomedia UK)*

Figure 3.15(ii)
Removing the second media tray screw (Tim Watson, Photomedia UK)

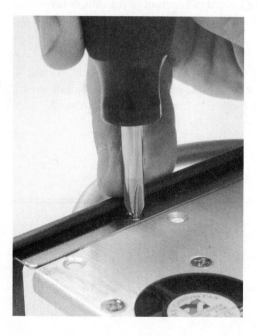

Figure 3.15(iii)
Removing the third media tray screw (Tim Watson, Photomedia UK)

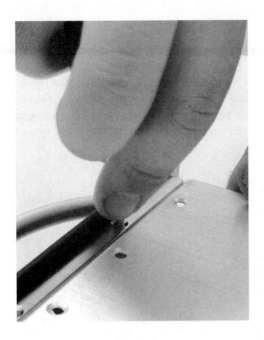

Figure 3.15(iv)
Removing the fourth media tray screw (Tim Watson, Photomedia UK)

Once this has been done you can put the screws in the pot, and put the media tray on one side until a little bit later. You will be left with two parts which look something like Figure 3.16.

Figure 3.16
Media tray separated from chassis (Tim Watson, Photomedia UK)

We are now going to work on the main chassis which is the part on the left of Figure 3.16. We will be installing the motherboard in this part of the case first, followed by fitting the drives to the media drive tray.

Before we can install the motherboard, we need to sort out the correct bezel for the rear of the case. Look in the Travla box, and you will find you are presented with an array of bezels (Figure 3.17). You will notice that not all of them fit your board. If you are not intending to fit any other motherboards in the Travla case, you can throw away all the bezels that do not fit your board.

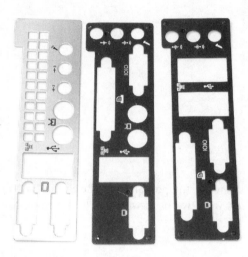

Figure 3.17
Array of bezels provided with the Travla case (Tim Watson, Photomedia UK)

For the EPIA SP motherboard, Figure 3.18 shows the correct bezel alongside the bezel supplied with the motherboard itself. The position of the holes for the ports on the bezel is similar in every respect, apart from the fact that there is no keyboard and mouse port on the bezel. The bezel is cut short because the keyboard and mouse ports are provided on the main chassis.

You will not be needing the bezel supplied with the Mini Itx motherboard; this can be discarded or retained for later use. However, keep it for a few moments to enable you to select the right bezel.

Now trial-fit the bezel against the motherboard. If you have picked the right one it should line up perfectly.

Now, holding the bezel onto the motherboard, lower the motherboard into the case carefully, seating the bezel against the rear of the chassis, and the motherboard against the four pillars that should locate near the motherboard screw holes.

You will find that the rear connector panel is sprung a little, so you will need to push the motherboard slightly toward the back of the case to align the screw holes with the chassis.

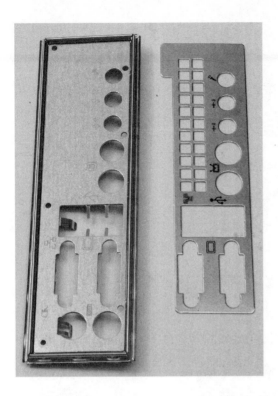

Figure 3.18
The right bezel next to the VIA original bezel (Tim Watson, Photomedia UK)

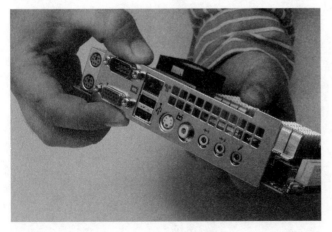

Figure 3.19
Trial-fit the bezel to the motherboard (Tim Watson, Photomedia UK)

Figure 3.20(i) to (vii)
*Fit the motherboard
into the case
(Tim Watson,
Photomedia UK)*

(i)

(ii)

(iii)

(iv)

(v)

(vi)

Figure 3.20 (Continued)

(vii)

You will now need to have a rummage in the little hardware pack that came with the Travla case and locate four coarse thread screws. These are the largish ones with the Phillips head. You will now need to screw the motherboard to the chassis. The screws should locate easily and be done up tight but not too tightly. Three of the screws are relatively easy to insert; however, there is one screw that goes in the back left corner of the case by the power connector. You must be careful not to break either of the metal tabs of this power connector, or to bend them in such a way that they would touch.

Figure 3.21(i)
Motherboard screw 1 (Tim Watson, Photomedia UK)

Figure 3.21(ii)
Motherboard screw 2 (Tim Watson, Photomedia UK)

Figure 3.21(iii)
Motherboard screw 3 (Tim Watson, Photomedia UK)

Figure 3.21(iv)
Motherboard screw 4 (Tim Watson, Photomedia UK)

With the motherboard screwed firmly in place, we need to install the memory. If you skimmed through Chapter 2, you may have missed one important point, although I will say it again here at the risk of repetition.

The memory for the Travla case *must* be low profile, that is to say "short" memory. The height of the memory module (i.e. the shortest length) must not be in excess of 0.8" or for those working in metric 20 mm. If you use larger memory, it will not clear the drives in the media tray above and you could damage your motherboard.

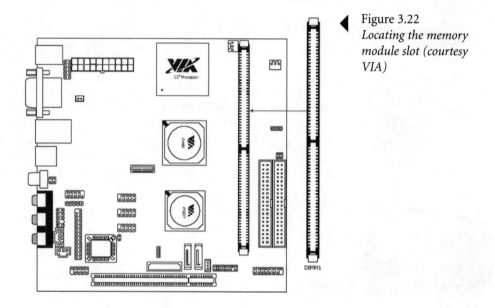

Figure 3.22
*Locating the memory
module slot (courtesy
VIA)*

To insert the memory module, line up the notches with the plastic moldings on the memory module slot (see Figure 3.23) and with a firm action, press the module home: the little white clips at the side of the motherboard should spring into position and lock the memory in place.

The next thing that we are going to do is connect the power supply. To understand how the Travla power system works, you have to realize that it has several components. The external power supply is the black plastic box which connects to the Travla chassis. We are going to use this when we are running the Car PC from the mains. It converts our line voltage to something suitable for our Car PC to handle, similar to our car battery voltage. The internal PSU is a DC-DC convertor that takes this 12 V nominal feed and converts it to the different voltages the motherboard requires to operate. Primarily 12 V, 5 V and 3.3 V.

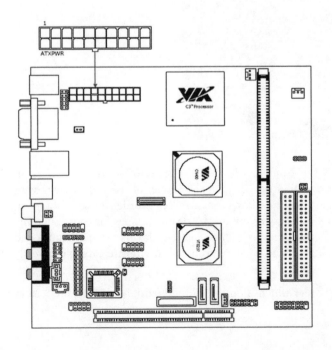

Figure 3.23
Align the notch in your memory with the plastic protrusion in the socket.

The internal power supply connects to the power jack at the rear of the case and accepts an input from either the mains adaptor or a stable supply from the car. We need to connect the ATX power cable from the power supply to our motherboard.

Figure 3.24 shows the location of the ATX power connector on the motherboard.

Figure 3.24
Location of ATX power connector on EPIA SP motherboard (courtesy VIA)

You simply need to connect the connector as shown in Figure 3.25. The ATX cable is quite bulky, and it can present a bit of a problem when putting the lid on the case. You might like to think of securing it neatly with some zippy straps or cable ties.

Figure 3.25(i) and (ii) *Connecting the PSU (Tim Watson, Photomedia UK)*

(i)

(ii)

Next we are going to connect the front panel Firewire connector – also known as an IEEE1394 port or DV port for digital video. There are a number of uses for Firewire with your Car PC. You might like to download video from a DV camcorder for safe storage to your car if you are on a driving holiday, freeing up your camera tapes for more shooting action. Or you might want a fast external hard drive interface for additional *.mp3s and movies. Whatever way you look at it, and whatever you call it, Firewire capability greatly enhances your Car PC.

The front panel connects to a header which is on the motherboard. Look at Figure 3.26 to locate the header.

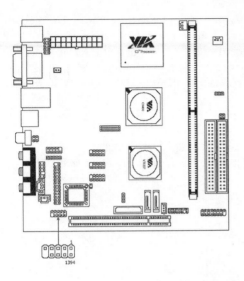

Figure 3.26
*Location of internal Firewire header
(courtesy VIA)*

Connecting the Firewire is simply a matter of locating the correct plug leading from the front panel printed circuit board, and connecting it to the header, noting the orientation of the plug in one of the pinholes that prevents incorrect insertion. The plug should connect to the header quite simply, no force should be required.

Figure 3.27
*Connecting the front panel Firewire to
the internal header (Tim Watson,
Photomedia UK)*

Next up is the front panel USB which has a very similar header to the Firewire. The motherboard comes with three USB headers, which is really useful considering the demand for USB ports in a Car PC application. The motherboard presents two ports at the rear panel and another two at the front, but there is nothing to stop the industrious Car PC builder from extending these headers outside the box and connecting them to USB socket PCBs that are not part of the Travla case.

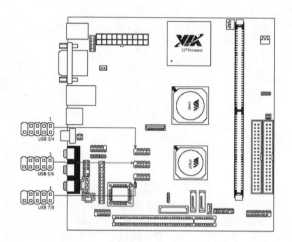

◀ Figure 3.28
Locations of internal USB headers (courtesy VIA)

Again, making the connection itself is relatively easy as long as you identify the correct plug for the header. A picture of the connection being made is shown in Figure 3.29.

◀ Figure 3.29
Connecting the front panel USB to the internal header (Tim Watson, Photomedia UK)

You need to think carefully about whether you want to connect the front panel audio connectors. Although it can sometimes be handy to access connection from the front of the case – especially when the unit is installed in a car, you must weigh up the benefits of having front audio access against losing the rear audio port. You can only select one option – front mounted audio or rear mounted audio. This is done using jumpers on the motherboard.

I strongly suggest that if you are planning on installing this unit into your car you consider using the rear audio connectors; this way the cabling to your amplifier can be concealed out of sight.

NOTE

If you want to use the rear audio connector, pins 5&6 and pins 9&10 must be connected or else the rear audio socket will not work.

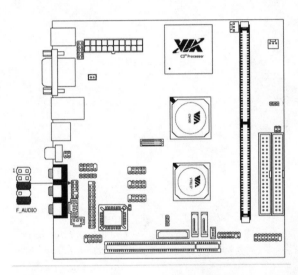

◄ Figure 3.30
Diagram showing position of front panel audio connector (courtesy VIA)

Pin	Signal	Description
1	FRNMIC	Front panel microphone
2	AGND	Ground used by analog audio circuit
3	MIC_BIAS	Microphone power
4	+5V AUDIO	VCC used by analog audio circuit
5	LINE_OUT_R	Right channel audio signal
6	NEXT_R	Right channel audio signal return
7	NC	No connection
8	Key	No pin
9	LINE_OUT_L	Left channel audio signal
10	NEXT_L	Left channel audio signal return

▲ Figure 3.31
Front panel audio connector pinout (courtesy VIA)

Figure 3.32
*The front panel audio headers/jumpers
(Tim Watson, Photomedia UK)*

Figure 3.33
*The microphone socket
connection (Tim
Watson, Photomedia
UK)*

If you are going to be using a headset for speech recognition, then you may find the front mounted microphone socket very useful as it allows easy connection and disconnection of a microphone.

To connect the front panel microphone socket to the motherboard, you will need to connect the "Mic-In" to "Pin1" of the F_Audio connector, "Ground" to "Pin 2" of the F_Audio Connector and "Mic-Pwr to Pin 3."

Figure 3.34
The lineout socket connection (Tim Watson, Photomedia UK)

IDEA

You might like to think of other uses for these front mounted connectors. In my book *50 Awesome Auto Projects for the Evil Genius* I show you how you can construct a simple circuit that allows infra-red communication between your Mini Itx motherboard and other devices. If you mount the photodiodes in a similar package to the "Total Remote" (described later in the book) you can make a neat infra-red transceiver that plugs into the front of your Car PC.

The front-mounted lineout connection is probably not as useful to you if you are mounting the Car PC in the front of your vehicle and connecting it to your vehicle speaker system. However, if you are building a Car PC for passenger use only, say in the rear of your vehicle, then the front mounted line out port is great for headphone connection to the motherboard. This allows the rear seat passengers to enjoy a game or a movie, without distracting the driver.

Figure 3.35 shows the location of the front panel header on the motherboard.

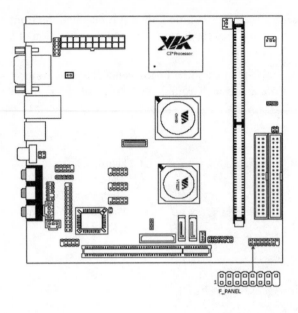

◀ Figure 3.35
*Diagram showing the location
of the front panel header
(courtesy VIA)*

◀ Figure 3.36
The front panel header

◀ Figure 3.37
The front panel connections (Tim Watson, Photomedia UK)

We are going to move on now to preparing the media tray.

▼ Figure 3.38(i)
Laptop media drive top (Tim Watson, Photomedia UK)

▼ Figure 3.38(ii)
Laptop media drive bottom (Tim Watson, Photomedia UK)

So here is our laptop media drive.

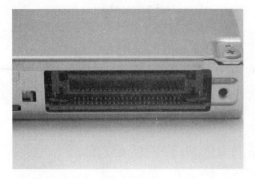

Figure 3.39
Laptop media drive IDE and power connection (Tim Watson, Photomedia UK)

Laptop media drives have a vastly different connection to desktop IDE drives. There is a single connection to the drive which integrates power, data and sound functions. To connect the drive to our Car PC, you will need to purchase an adaptor. Digital WW sells one such adaptor. It appears on their website as "Slim to IDE Adaptor" with the reference SLIMIDE.

W³ www.digitalww.com

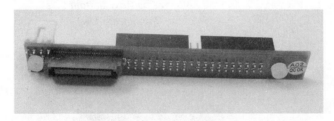

Figure 3.40(i)
Laptop media drive adaptor front (Tim Watson, Photomedia UK)

Figure 3.40(ii)
Laptop media drive adaptor back (Tim Watson, Photomedia UK)

The adaptor converts the 50 pin laptop media drive IDE interface, to a standard 40 pin IDE interface, power connector and pair of audio connectors.

The power connector is different from that which would be found on a desktop $5\frac{1}{4}$" media drive. The connector is of the variety more normally found on smaller $3\frac{1}{2}$" drives like floppy drives. The Travla power supply provides a connector for the media drive, but we will come to this later.

The audio connector is to allow analog audio to connect directly to the motherboard. You will need an additional cable for this – in fact it is the same cable you would use in a conventional PC. You will not find that you need the same length cable as in a PC, so either cut an existing cable to make it shorter and insulate the connections with heat-shrink, or, if you can, buy a smaller cable.

Figure 3.41
Attaching laptop media drive IDE adaptor to the back (Tim Watson, Photomedia UK)

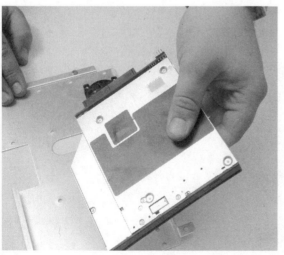

Figure 3.42(i) to (v)
Locating the laptop media drive (Tim Watson, Photomedia UK)

(i)

Figure 3.42 (Continued)

(ii)

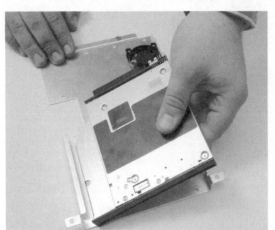

(iii)

(iv)

(v)

You now need to look in the hardware package that comes supplied with the Travla case, and identify the two really small screws. They are pictured here next to a pound coin. You will certainly need a jeweler's screwdriver, and you may find that you need to use a pair of tweezers to hold these screws. The screws are used to secure the laptop media drive.

◀ Figure 3.43
Tiny screws next to pound coin
(Tim Watson, Photomedia
UK)

The screws secure the drive on the side opposite the channel where you located the drive. There are no screws on the other side where the channel is located, or anywhere else on the drive. Secure the two screws, one at the front, and one at the back, to prevent any movement of the drive. Figure 3.44(i), (ii) shows the location of the two screws.

Before proceeding any further, I can wholeheartedly recommend reassembling the unit, screwing the media tray to the inside case and then fitting the outside case on briefly to ensure that the media drive lines up correctly with the outside of the enclosure. The difference between a great installation and a mediocre one is in the detail. By ensuring that the small things are right you will make the installation much better.

Figure 3.44(i)
Laptop media drive screw 1 (Tim Watson, Photomedia UK)

Figure 3.44(ii)
Laptop media drive screw 2 (Tim Watson, Photomedia UK)

Figure 3.45
Laptop media DVD super drive installed (Tim Watson, Photomedia UK)

Once you have satisfied yourself that the drive is flush with the outside of the case, tighten the screws and possibly make a mark somewhere inconspicuous with a permanent marker, that will allow you to line up the media drive with the disk tray if you need to reassemble it.

Figure 3.46
Front view of case with drive installed (Tim Watson, Photomedia UK)

Pay attention to the orientation of the IDE cable when inserting it into the adaptor. There may be a plastic protrusion on the connector body that only allows one way insertion; alternatively, one of the holes in the middle of the connector may be blanked off allowing insertion in only one orientation.

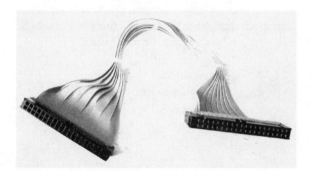

Figure 3.47
Media drive IDE cable (Tim Watson, Photomedia UK)

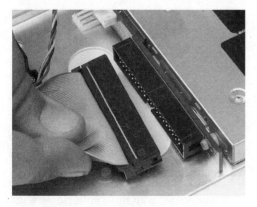

Figure 3.48
Attaching the laptop IDE cable (Tim Watson, Photomedia UK)

▼ Figure 3.49(i)
Endurastar front (Tim Watson,
Photomedia UK)

▼ Figure 3.49(ii)
Endurastar rear (Tim Watson,
Photomedia UK)

Your drive may present you with a selection of screw holes for mounting your drive. In some models you can screw the drive to the carrier tray from both sides of the drive. If this is the case, you need to scour your drive for a small label that advises "Do not cover this hole" (see Figure 3.50 for the label on the Endurastar drive). It is imperative that this hole is kept clear so that the hard disk can "breathe." When mounting inside the Car PC, this label should be pointing downwards towards the motherboard and not against the metal plate.

◀ Figure 3.50
Heed this warning (Tim Watson,
Photomedia UK)

Another thing you will need to check, especially if you are using a second-hand hard drive, is that the drive is set correctly to either "Master" or "Slave." On our Hitachi drive, the unit is supplied set to "Master" as standard from the factory. Look at Figure 3.51. You are looking for a similar label on your drive that shows you how the jumper should be set. We want to set our drive to "Master." Looking at the label, we are presented with three options, we want the first. It should be apparent that the little circles on the label correspond to the pins poking out of one end of the drive. If we look at the label, we can see that there is a group of four pins, and a group of two set to one side.

It is the four pins that we are interested in, the two pins that are set to one side give us the visual cue as to what way the four pins are oriented. They are representative of the rest of the laptop IDE connector.

◀ Figure 3.51
Endurastar jumper legend (Tim Watson, Photomedia UK)

Before we screw the drive down to the tray, we need to inspect the drive carefully for any indication of the orientation of the laptop IDE connector.

On the Hitachi Endurastar drive, there is a "1" on the printed circuit board legend nearest to the "jumper pins" we mentioned in the last bit. We need to make a mental note of this – or even better write it down.

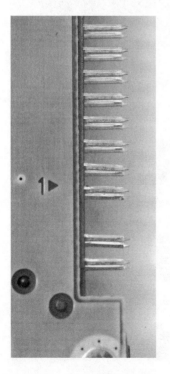

◀ Figure 3.52
Endurastar: Note the position of the "1" (Tim Watson, Photomedia UK)

The next step is mounting the drive. Go through the hardware package to find four of the small countersink bolts with the fine thread.

When using the Endurastar drive, you will only need to use four of the six holes. Align the drive so that it sits correctly on the media tray, and while supporting the drive with one hand, tighten the four bolts to secure the drive. Figures 3.53 (i)–(iv) show the location of the four screws.

▼   Figure 3.53(i)
Hard drive screw 1 (Tim Watson, Photomedia UK)

▼   Figure 3.53(ii)
Hard drive screw 2 (Tim Watson, Photomedia UK)

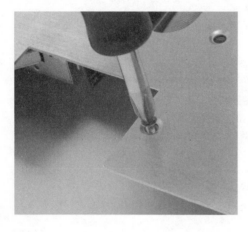

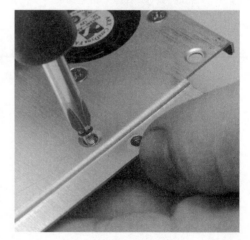

▼   Figure 3.53(iii)
Hard drive screw 3 (Tim Watson, Photomedia UK)

▼   Figure 3.53(iv)
Hard drive screw 4 (Tim Watson, Photomedia UK)

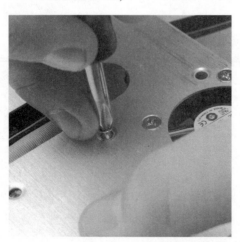

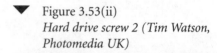

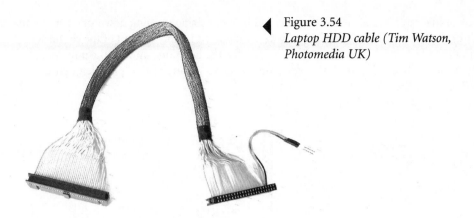

Figure 3.54
*Laptop HDD cable (Tim Watson,
Photomedia UK)*

Figure 3.55
*Connecting the hard drive IDE cable
(Tim Watson, Photomedia UK)*

Remember when you were installing the hard drive and I asked you to note the position of the number "1," Well, you're going to need to have that information at hand. If you have bought wisely, and are working with the Endurastar hard drive, I'll remind you that the "1" was nearest to the block of four jumper connections to one side. If you have another drive, this may not be the case.

You need to take the remaining IDE cable supplied with the Travla case. This will have a standard 40 pin IDE connector on one end; on the other is a laptop hard drive 44 pin connector with a short spur with a couple of wires attached and a small white plug.

Look at the gray ribbon cable – there should be a red stripe running down one wire to one side of the ribbon cable. This denotes "wire 1." You need to line this end of the connector up with the end where you saw the "1." Make sure that the holes line up with the pins, and press the connector firmly home all the way until you can't see any of the "pins" left between the hard disk and the connector.

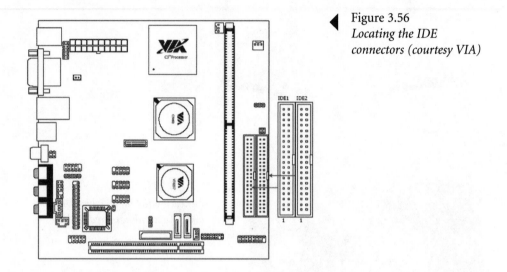

Figure 3.56
Locating the IDE connectors (courtesy VIA)

As is mentioned earlier in this book, I selected the Hitachi Endurastar hard drive on the grounds that it is a very durable drive with a built-in shock sensor, designed to perform well in extreme environments.

As an alternative, the Travelstar E7K100 is a worthy adversary to the Endurastar. It boasts a massive 100Gb for media junkies and will be able to take advantage of the blazingly fast SATA interface on the VIA EPIA SP motherboard. One disadvantage is that this drive is not engineered for the harsh environments that the Endurastar was destined for and therefore may not last as long, especially if you are anticipating driving in extreme conditions.

If you decide to use a SATA Travelstar, I recommend that you find the shortest SATA lead possible as space within the case will be very limited when it is all assembled. For reference the SATA connections are located next to the BIOS backup battery. This is illustrated in Figure 3.57.

Figure 3.58 shows what it should all look like if you have done everything properly.

Alternatives to a hard disk

The Hitachi Endurastar is a *very* rugged drive, and should cope admirably with even the most demanding conditions. However, some builders of Car PCs would prefer to experiment with Solid State storage. This eliminates any moving parts that are present in a hard disk.

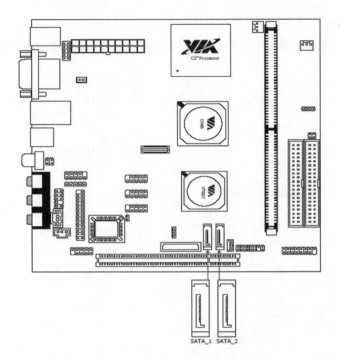

Figure 3.57
Locating the SATA connectors (courtesy VIA)

Figure 3.58
The completed media tray (Tim Watson, Photomedia UK)

In their favor is that the devices are all based around chips, with no moving parts to wear out and degrade; the devices also have much lower power consumption than a hard disk. Against them is the fact that flash memory can only be written to a certain number of times, and as a result the applications are limited. In addition, solid-state storage is much more expensive per gigabyte than an equivalent hard drive.

You will be familiar with Compact Flash cards; they are commonly used in digital cameras to store information. We can also use them as a "hard disk substitute" in our Car PC. Some motherboards such as the VIA MII, will readily accept a Compact Flash Card. The VIA EPIA Mini ITX SP, however, requires an IDE to CF adaptor (see Figure 3.59).

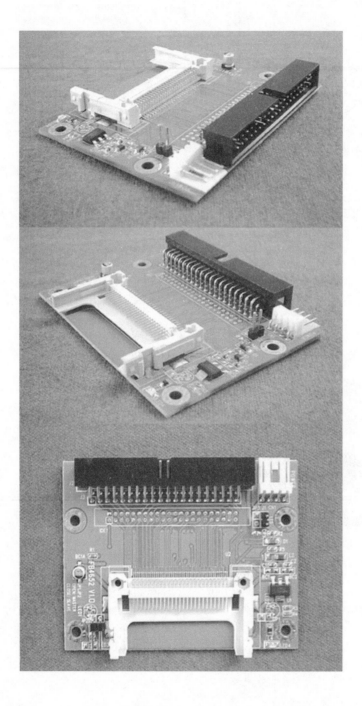

Figure 3.59
*IDE to CF adaptor
(courtesy Travla)*

Travla supply these, although they do not supply a variant recommended for our choice of case. It is, however, possible (with a little persuasion and modification), to mount an IDE to CF adaptor containing a Compact Flash card in place of a hard disk. This will severely limit your storage; but it can provide a more stable solution for the most rugged of applications.

The hardware installation requires that you mount the Compact Flash card (see Figure 3.60) in the adaptor.

Once this is done, the adaptor can be connected to your Car PC as an ordinary IDE device.

Setting up the software on a Compact Flash card is harder than with a conventional installation. You need to download Windows XP Embedded from Microsoft – they offer a free trial:

◀ Figure 3.60
*Compact Flash card
(courtesy Crucial)*

http://msdn.microsoft.com/embedded/windowsxpembedded/default.aspx

There are details on the Microsoft website about installing Windows XP Embedded and using the Microsoft Development Tools.

You will need to set up an Enhanced Write Filter, essentially this "filters out" all non-essential writes to your Compact Flash card, to try to keep the card in tip-top condition (they are not too keen on being written to too many times).

You can go to the Microsoft website at:

http://msdn.microsoft.com/library/default.asp?url=/library/en-us/dnxpesp1/html/ewf_winxp.asp

for full instructions on how to set up an Enhanced Write Filter.

Connecting the media tray to the main board

Now comes a stage when you may require an extra pair of hands – mating the media tray to the main motherboard, and then screwing it to the chassis.

There are a number of connections that need to be made between the media tray and the motherboard. Unfortunately, because of the space constraints of the design, cable lengths have been kept to a minimum which necessitates holding the media tray at right angles to the chassis, while connecting the headers and connectors in close proximity. The cables must then be found appropriate positions, and the cables connected to the motherboard.

Figure 3.61
Connecting the media tray fan connector (Tim Watson, Photomedia UK)

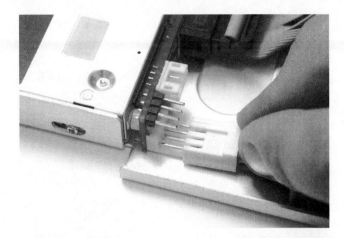

Figure 3.62
Connecting power to the media drive (Tim Watson, Photomedia UK)

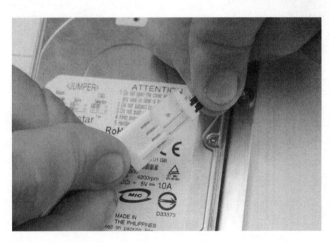

Figure 3.63
Connecting the hard drive power connection (Tim Watson, Photomedia UK)

Chapter 4

Installing the operating system, motherboard drivers and choosing a front end

Right – you've finished the hardware, and your Car PC is looking great. You now need to make it do something useful, because if you plug it in and switch it on at the moment, about the most meaningful thing that it is going to do is give you a non-system disk error.

> **Hint**
>
> For the following steps, get hold of a bog standard CRT monitor, and a keyboard and mouse. You will find it MANY MANY MANY times easier doing the installation using plain old vanilla hardware than with your fancy Car PC interface devices which probably (more or less definitely) won't work at this stage without a GUI and drivers.

The same as with any other PC, you will need an operating system.

The alternative to Windows (some would say the dark side...) – Open Source

There are lots of people out there on the web that extol the virtues of the "Open Source" movement. There is a wealth of free software available and a very active community support Linux amongst other "free" operating systems. If you feel competent, there is no reason why you cannot install Open Source software on your Car PC: the advantages are clear, it is free software developed by the community for the community with no commercial interests. Despite its many advantages, many users are not familiar with anything other than a bog standard Windows interface, so here we are going to cover the basics, installing Windows XP™ (see Figure 4.1) a product that most people are familiar with. For those who are slightly more intrepid, maybe familiar with Linux and open source platforms, see the box "Open Source Resources" for a list of links to Car PC resources.

OPEN SOURCE RESOURCES

Here are links to websites where you can read more about Open Source software and download some free software for your Car PC.

Debian Linux	W³	www.debian.org
Fedora Project	W³	http://fedora.redhat.com/
Free BSD	W³	www.freebsd.org
Gentoo Linux	W³	www.gentoo.org
Knoppix	W³	www.knoppix.com
Suse Linux	W³	www.suse.com
Ubuntu Linux	W³	www.ubuntu.org
Xandros Desktop OS	W³	www.xandros.com

Installing Windows XP

Figure 4.1
Windows XP Home box shot (courtesy Microsoft)

Windows XP Home should be fine for the majority of users. Windows XP Professional does offer a few additional "Power Features," but for a basic Car PC setup XP Home is fine.

Some of the key differences between the two packages are:

- An XP Home Machine is unable to become part of a Windows Domain; this is only really applicable to users who want to integrate their Car PC with an advanced network – for most home setups this will not be an issue.
- XP Home Machines do not have the "Remote Desktop" functionality built into Windows XP Professional that allows a user to control your PC remotely – in some ways this is a blessing as a number of users have reported security problems when this feature is inadvertently enabled.
- Windows XP Home does not have encryption built into the software.
- Windows XP Home does not allow allocation of file permissions on an individual basis.

Like I said, for most users this will not be an issue.

NOTE

Before thinking about installing XP, make sure you have NO USB devices connected, not even a keyboard or mouse.

So now you need to tell your Car PC that it is booting from CD, in this case our laptop media drive. To do this we need to go into the BIOS. In times gone by, much had to be configured manually, however, modern BIOS's seem to do a pretty good job of setting things up automatically, and certainly with the setup we have here there should be no need for manual intervention.

▼   Figure 4.2
    BIOS set to boot from CD (courtesy Microsoft)

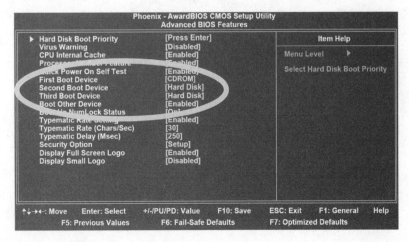

Now when you switch the Car PC on you should insert your Windows XP CD into the media drive at the front of the case. If you bought the one I told you to it will be sucked into the slot gracefully with all the nuances of a high-end piece of in-car entertainment.

Unfortunately, for the time being, this illusion is quickly spoiled by the "80 character" text only DOS display.

If you now boot from the CD, you will be greeted with the welcoming message:

> "Setup is now inspecting your computer's hardware and configuration…"

In the first section of the Windows XP installation, you will see an MS-DOS based screen, asking you to press F6 if you want to install any third party or RAID drivers. In this case we don't, so don't press a thing! There really isn't room in our 1 DIN case for a RAID array.

Before we can attempt to install Windows, we need to format and partition the hard disk. This is basically checking that the disk is OK, smoothing out the wrinkles, and setting up a file system that Windows can use. I strongly recommend that you format the whole lot as a single NTFS partition unless you have a good reason to do otherwise. Luckily, with the Windows XP install, Windows now takes care of formatting the drives: all you have to do is tell the computer how many "pieces" you want the drive to be split into and how big each "piece" should be. In fact, the correct name for these "pieces" is partitions. You will see the screen in Figure 4.3

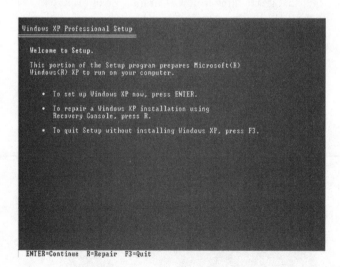

Figure 4.3
Windows XP
Installation Step 1
(courtesy Microsoft)

You will now need to press ENTER, as we are starting a new, fresh installation. You will then be asked what partition you want to format. You will generally have a number of options here, but go for the one with the largest "MB" number. This should be in the order of thousands, 30-odd thousand if you picked the Hitachi Endurastar drive that I recommended in Chapter 2.

We now move on. Refer to Figure 4.4 for the next step of the installation.

▼ Figure 4.4
Windows XP Installation Step 2 (courtesy Microsoft)

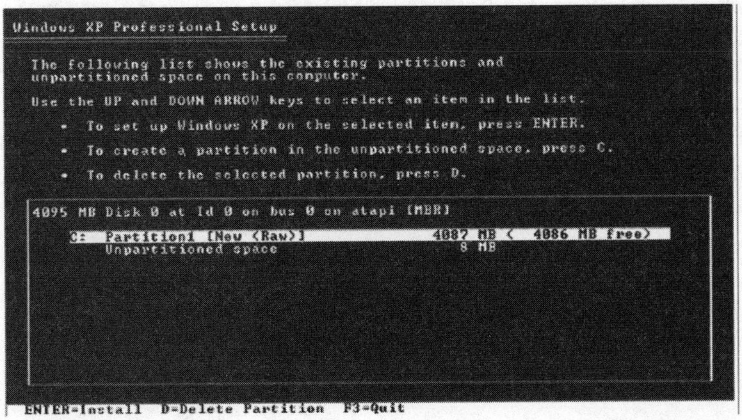

You now need to tell the computer to do a FULL format, not a quick format, and make sure the disk is formatted as NTFS. It takes a little more time, but it is best to start off on the right foot by doing things properly. NTFS is the newer, enhanced file system used by Windows XP and some other versions of windows (versions based on the Windows NT "backbone"). It is vastly superior to FAT (File Allocation Table – the file system used on Windows 9x) in many respects.

You're now going to get a screen that looks like Figure 4.5

▼ Figure 4.5
Windows XP Installation Step 3 (courtesy Microsoft)

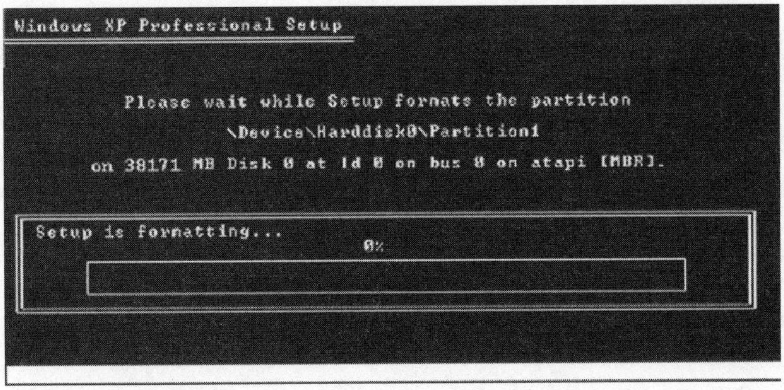

This is going to take some time. The amount of time is determined by the size of the disk and the speed of the processor.

Hint

Now is a great time to feed the cat, turn the cooker off, make some coffee, grab a beer or do whatever else your partner asked you to do last night before you sat down to watch the game. Formatting the hard disk is going to take a while.

That's the boring bit done. Windows should carry on for a while on its own, sorting things out until it needs a bit of information from you. So go on, spend some quality time with your family for the next few minutes.

The first thing that Windows will ask you for is the Windows XP product key. This is to check that you are using a genuine copy of Windows XP, and not a dodgy CD from Tony at the fleamarket.

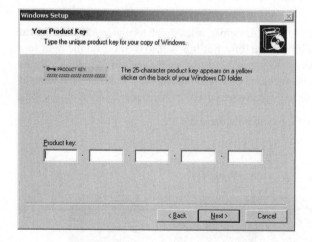

Figure 4.6
*Windows XP Installation
Step 4 (courtesy Microsoft)*

You are looking for a yellow sticker on the back of your Windows CD case. Enter a valid key, and Windows will pick up from where it left off, doing everything automatically for a little while. The screen will change momentarily from one item to another, and the little green bar will crawl across. Don't worry! Even if your PC doesn't appear to be doing much, Windows XP is being installed behind the scenes.

The next time Windows stops is to ask for your regional setting and text input devices. This is the language that you will be using Windows in and the hardware you will be using to input text.

Figure 4.7
Windows XP Installation
Step 5 (courtesy Microsoft)

Hint

Make sure you set your keyboard to UK if it is a UK keyboard, otherwise you will find lots of really little irritating details like the " and @ symbols being reversed – to mention but one.

Figure 4.8
Windows XP Installation
Step 6 (courtesy Microsoft)

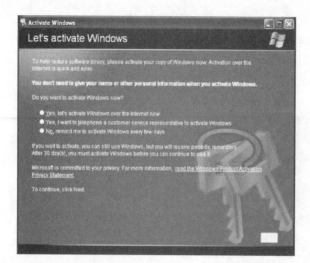

Figure 4.9
Windows XP Installation Step 7 (courtesy Microsoft)

Once you have finished this step, you will be brought back to the main installation procedure, which will now ask you to register the product. Be sure to have the information supplied with your Windows XP disks. You will need to work through the procedure which will register the product with Microsoft and authenticate it for use. Once this is complete, Windows XP can be considered "installed." You now need to proceed to installing the drivers.

The drivers for the EPIA motherboard come supplied on a CD. Alternatively, if you have bought your motherboard second-hand and did not get a driver disk supplied, you might find www.via.com.tw throws up what you are looking for. (Incidentally, you can also download Mini ITX manuals from the same site.)

When you are installing the drivers, you must insert the disk into your Car PC media drive. You will be presented with the menu shown in Figures 4.10 and 4.11. You need to select all

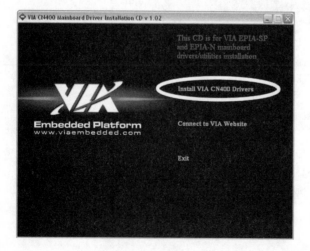

Figure 4.10
Driver installation Step 1 (courtesy VIA)

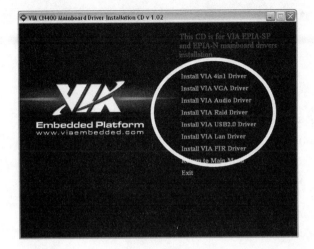

Figure 4.11
*Driver installation Step 2
(courtesy VIA)*

of the options enclosed by the white line one by one. Complete installation of one driver before proceeding to the next.

Another thing I would *highly* recommend you to do is to download the latest service pack for Windows XP if you are using an older disk. This can be downloaded free from Microsoft's website at:

W[3] www.microsoft.com/windowsxp/sp2/default.mspx

One of the main issues with Car PCs is that they are not "instantly on." When you turn on your car stereo, it is instantly working, the radio comes on when you press "on" – you don't need to wait for it to boot.

One of the ways that we can get around this is by putting our Car PC into an energy saving mode. Rather than shutting the Car PC down each time which would require a full reboot, we can turn off the stuff we don't need and just keep critical information in the Car PC's memory. A little like being asleep, the information is still there, but the Car PC is unable to react to external stimulus.

One of the things that we can do, is tell our Car PC to go into "hibernation" every time the power button is pressed. This is relatively simple. We need to go into "Control Panel" followed by clicking on "Power Options." Once we have done this, we need to click on the "Advanced" tab, and from the drop down menu of "When I Press The Power Button of My Computer" select "Hibernate."

What you have done may be enough to get you up and going. However, XP is quite a chunky bit of software, and you may find that it runs a little slowly compared to your home PC. What you can do is "tweak" windows XP to turn off the things that you really don't need. If you are an advanced user and you are feeling brave, try this link:

W[3] http://www.tweakhound.com/xp/xptweaks/supertweaks1.htm

This will give you an in-depth guide.

Choosing a Car PC front end

Autotouch

 http://www.sarinarts.com/autotouch.htm

Autotouch is a relatively simple piece of Car PC front end software. It relies upon Windows Media Player 9.0 for its multimedia functionality and MapPoint for its GPS functions. It is written in C# and requires the Microsoft .NET framework to be installed to function.

AutoPlay

 http://autoplay.magnetikonline.com/

AutoPlay has not been supported since 2003; however, for basic functionality it is still a useful piece of front end software and available from the net as freeware.

- Mp3 support is provided courtesy of Winamp
- Autoplay will play Divx movies when the codec is installed
- Supports infra-red remote control with IRman
- Designed for low-res composite 320×240 car monitors
- Skinnable

Centrafuse

 www.fluxmedia.net

Centrafuse is one of the few pieces of Car PC software that offers multilanguage support.

One of the things that I like about Centrafuse is its support for Destinator 3 as a GPS program, I believe that Destinator is one of the best GPS programs on the market. The On-Screen keyboard built into Centrafuse is nice and clear and allows easy entry, and it is certainly much nicer than the built-in OSK in Windows.

Centrafuse requires that you install the .NET framework SAPI (for speech recognition) and Direct X 9.

You can currently download the Beta 1.5 version from:

 http://www.centrafuse.com/downloads.aspx

Some of its more salient feature include:

- Audio mixer controls
- Three screen dimming modes
- Customizable "hot key" support
- Id3-tag and directory mode
- Integrated playlists and favorites
- Integrated fm radio support (with radiator)
- Video playback (poster image support in video manager)
- Integrated dvd
- Internet status (lan,gprs,wireless)
- Ability to disconnect/connect default internet connection
- Integrated weather and web browser
- Free DB with a local DB for storage (used when CD inserted and ripping)
- Speech/voice recognition
- CD ripping
- Fully embedded GPS support
- Fully integrated phonecontrol .NET

Media Car

Media Car hails from Belgium and is considered no longer supported. It is still a funky looking piece of front end software with support for Mp3, radio (with a D-link USB radio), GPS and Divx movies.

One feature which looks great when fired up, but isn't especially useful is the "Mixer" function, which presents the Winamp Sound Controls as a series of sliders, on a screen allowing you to control them with the touchscreen. While this looks pretty, software equalizers are not a substitute for pro-quality hard-wired equalizers, but with the AC'97 audio of the EPIA SP taken into consideration, this is not a bad function.

One thing which may appeal to power users is the fact that Media Car supports text LCDs.

Mobile Impact

 http://mobileimpact.biz.tm/

Mobile Impact is a Car PC front end based on Media Player 9. It boasts integrated XM radio support and phone-control software and unusually features a game emulator that plays Nintendo and Sega games. It incorporates GPS support for iGuidance or Freedrive.

Mobile Media Center

- Support NTSC mode 352×240, 352×480, 720×480 and PAL mode 352×288, 352×576, 720×576
- Support 24 fps or 30 fps DVD movie
- Support 4:3 and 16:9 aspect ratio
- Support 32, 44.1 and 48KHz sample rate for AC-3 Audio
- Support mix 5.1 channel into 2 channel Dolby surround stereo sound
- Support sample rate 48KHz and 96KHz for Linear PCM (LPCM)
- Support 16, 20 and 24 bits/sample
- Support MPEG-1 and MPEG-2 for MPEG Audio
- With up to 32 subtitles, 52K biggest data pack and highlight button
- Support up to eight languages
- Support point to point loop function
- Support LINE21 output
- Video lightness and color adjustable
- Support up to eight parent control grades
- Multiple view angles, up to nine view angles
- Support up to 32 subtitles
- High performance de-interleave technology

Chapter 5

Connecting your Car PC
to the "Real World"

In this chapter, we are going to talk about installing your Car PC and some of the issues that you are going to come across. You are going to need to do a little bit more work if you want your Car PC to work "in-car". To carry out the procedures in Chapter 4 you used a standard keyboard mouse and monitor; however, for in-car use these peripherals are a little "clunky" and you will want something easier to use.

The chapter is split into roughly three parts:

- First of all, we are going to consider how to power your Car PC.
- This is followed by looking at display options for your Car PC.
- Lastly, we will be exploring a number of ways you can connect your Car PC's audio to your car speaker system.

Getting to grips with in-car power

So far you have been using your Car PC plugged into the mains running off line voltage with the Travla PSU pictured in Figure 5.1. Well, in the car things are slightly different. For a start, the power you are using is not unlimited. Whereas at home you can just suck a few more kilowatts out of the grid, in a car, you have a finite amount of power, and you have to maintain the balance between meeting your entertainment needs and providing enough power for vital functions, ignition, bodywork electronics, lights, instruments – the list goes on.

A very CLUMSY, and note the capitalization, way of powering your Car PC is to use an inverter. What this does is takes the *low voltage DC* emanating from your car's battery and turns it into a *high voltage (approximation of) AC*.

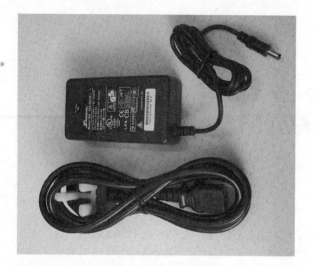

Figure 5.1
Travla PSU

Notice the word "approximation." The inverter doesn't actually produce AC it produces a choppy approximation of the wave form.

When we change the type of electricity from one form to another, we invariably lose a little as heat. The irony of this approach is that now we have this lovely approximation of line voltage, we take it and feed it through a transformer that changes it back to similar levels to our car battery. This then feeds into our Car PC which takes this (roughly 12 V) and turns it into regulated steady voltages the motherboard can deal with.

Thankfully as the VIA motherboards are so well designed they consume very little power, but the main caveat with this approach is the amount of power wasted.

The next thing that you can do is wire a 12 V cigarette lighter to a DC input jack plug lead. This does work (for a spell) although it is a bit of a dumb idea. Here is why. The internal PSU assumes that it is receiving a nice clean 12 V signal. The problem is, the voltage in your car *isn't* nice and clean. If you wire the 12 V from your car directly, your motherboard is liable to fail rather quickly, and the least sporadic operation will result.

The *proper way* to do things is to use a DC–DC converter. This allows your Car PC to survive when you start the engine, and it smoothes out all those little power fluctuations.

A good choice for the Travla C150 case is the CarNetix CNX-P1260 12V DC–DC regulator. This is well matched to the Travla case PSU's capabilities as it will provide up to 60 watts of power, which is the same rating as your Travla's internal PSU.

If you want to go this route you can download a connection diagram from:

W[3] http://www.carnetix.com/applications/connection_detail.htm

It requires connections between a permanent 12 V feed and a good earth as well as a connection to the ignition switch accessory circuit.

On the PC side of things, it provides a regulated 12 V to feed to the internal PSU, as we a connection to the "Front Panel Connector" (see Chapter 3 to refresh your memory). ' allows the power supply to start up and shut down your Car PC when you start your ig tion. Cool, huh? Additionally, another nice feature is that it provides an accessory outpul connect to things like amplifiers and electric aerials to energize them with the Car PC.

When connecting a Car PC, it is wise to consider how the additional load is going to be dealt with by your car's electrical system.

Generally manufacturers design everything with a pretty narrow margin. There is no point in putting in a 100 A alternator if they can get away with an 80 A. What this means is that your car may not generate sufficient power to allow you to run your Car PC permanently. If you only want to use it momentarily, then no worries; however, if you are looking at a permanent installation to use on a regular basis, then you might want to invest in an auto electrician to see the cost of upgrading your alternator and charging system.

Additionally, your car's battery capacity is another consideration. While it may be tempting just to fit and forget a bigger battery, consider the prospects of fitting an additional leisure battery solely for your Car PC and in-car entertainment.

Leisure batteries operate differently to car batteries. The battery in your vehicle is what is known as a "cyclical battery." What this means is that it will tolerate being topped up and drained, topped up and drained, in a constant cycle. By contrast, a leisure battery, also known as a deep cycle battery, is more tolerant of a long charge followed by a long discharge.

If you look at Figure 5.2, we have a simple split charging circuit employing a single relay. It allows your second deep cycle battery to charge only while the engine is running. This is then isolated from the car's electrical system when the engine is not running (and hence no power being generated).

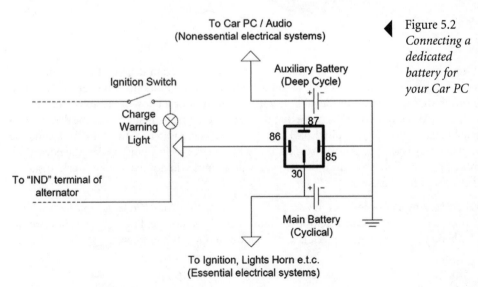

Figure 5.2
*Connecting a
dedicated
battery for
your Car PC*

This means that if you are using your Car PC while parked, there is no danger of you exhausting the charge in your car's main battery and as a result, you will not find your battery flat! Furthermore, by isolating your Car PC from your car's electrical system while the engine isn't running, you are isolating the Car PC from electrical noise present in the car's electrical system.

In-Car monitor options

Mounting styles

There are a number of ways to mount your monitor. Choosing a location for your monitor will largely depend on the applications that you want to use your Car PC for. A PC for satellite navigation will be largely driver centric, whereas an installation for multimedia will be more focused around the passengers of the vehicle, and may even be exclusively for rear seat passengers.

Goosenecks

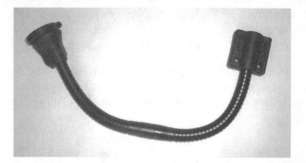

Figure 5.3
Gooseneck monitor mounts

A gooseneck (see Figure 5.3) is a mounting stalk with a certain degree of compliance, and a certain degree of stiffness, to allow you to bend and flex it and position your monitor in a position that is optimal for viewing. If you were to mount a gooseneck in the front of the vehicle, it would offer the advantage that the screen could be turned towards the driver when satellite navigation was required, and if a movie was being watched, the gooseneck could be angled towards the passenger instead. Clearly this degree of flexibility makes the gooseneck more practical than fixed mounting, however, goosenecks do not make for a "sleek" install as they are a little obtrusive.

When looking at goosenecks, try to find a model with a "quick release" mechanism that allows you to remove your monitor and stow it away for storage. Leaving a monitor prominently on display on a gooseneck is an invitation to opportunist thieves.

Headrest mounting

Integrating your monitor into your headrest is a great way of providing screens for rear-seat passengers. Obviously this would not be a good mounting method for a Car PC which was required to perform satellite navigations as it would be inaccessible by the driver! Despite this shortcoming, headrest mounting is ideal for multimedia installations, keeping the kids amused on long journeys for example.

There are a number ways of accomplishing headrest mounting. For a temporary installation, you might want to consider using Velcro or an elastic strap to secure the monitor to the headrest. This allows your Car PC to be moved between vehicles, and as nearly all cars have headrests, you are guaranteed easy mounting for your rear-seat passengers. The other option which is more permanent, is to reupholster your headrests, and remove some of the foam with a scalpel, allowing you to inset the monitor. You might have to take some of the rear seat squab of your seat apart in order to route the cables neatly and effectively.

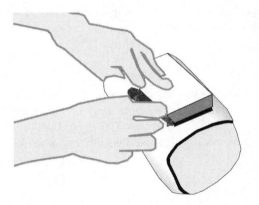

◀ Figure 5.4
Headrest mounting diagram

Other styles

Many monitors now come with a receptacle on the rear for a threaded bolt. In a number of cases, the threads are the same as that used for photographic equipment – a standard that you find on tripods. This opens up other mounting possibilities – nip into your nearest pro camera shop and have a look at the range of brackets, something suitable may be adapted to fit.

The Lilliput monitor shown in Figure 5.6 came with this natty plastic stand; this enables you to stick it to your dashboard using a self-adhesive base, and secure your monitor with the means of a screw on the back of the bracket.

While this may prove a simple option, it would be inadvisable to leave your monitor on display as it makes it a target for thieves.

◀ Figure 5.5
Stand for free-standing monitor mounting

If you jump ahead a little to the next few pages you will see some examples of monitors that have been designed to be mounted in a variety of situations:

Freestanding

These are the most versatile of monitors and lend themselves to many applications. Folks skilled in the art of fiberglass have successfully taken apart freestanding monitors and blended them into new fascia panels for their car using glassfiber mat. This is a real skill, and getting a nice result takes an element of craftsmanship which could not be recommended unless you are supremely confident.

In-dash

In-dash monitors take up very little space when retracted and look fairly inconspicuous when folded in making them less of a target to the opportunist thief. There are a number of manufacturers of in-dash screens, which mount in a standard 1 DIN slot in your dashboard. In many cases, an in-dash screen will provide an in-keeping replacement for an old car stereo.

Some of the monitors are manual requiring you to put the device out yourself. However, some of the better monitors are electric and fold out automatically. This is a really sleek feature which certainly looks very swish.

When positioning a fold-out monitor, you need to think carefully about what controls the monitor is going to obscure: make sure that there is nothing mission critical that you need to access while the monitor is open. Jump ahead to Figure 5.10 to see a nice flip-out monitor made by Digital WW.

In-roof

Monitors mounted in roof consoles are becoming a common sight. The monitor is supplied in a housing which mounts to your vehicle's roof. If you have a solid headlining made out of a hard plastic, then you may find the lining alone provides suitable mechanical support for your monitor. If you have a fabric headlining, and or your roof is only "single skinned," i.e. the only thing between the headlining and the sky is a single sheet of metal, then you may find it easier to mount a panel of plywood or similar, using contact adhesive, to the roof of your vehicle to provide something secure to screw into.

One of the advantages of mounting your monitor in the roof is that all cables can be run inconspicuously in A and B panels to your Car PC. Jumping ahead a little, Figure 5.9 shows a nice roof-mounted widescreen monitor made by Xenarc.

Monitor options

You've now built your Car PC and it looks great on the bench, however, you are going to want to use it in your car. Standard CRT and LCD monitors are way too bulky to fit inside a vehicle (unless you drive a truck), so let's look at some of the more compact options for your vehicle.

Connecting your Car PC to a VGA monitor

We are going to look at connecting your Car PC to a small LCD, VGA monitor. In this case, we are going to be using a 7" Widescreen Lilliput Monitor.

Your monitor will be supplied with a number of cables. We are going to take a moment to familiarize ourselves with these cables and accessories.

First you will come across a gray plastic "torpedo" on a length of wire. This will have a car cigarette lighter plug on one end, and a small PSU plug on the other. This is our 12 V power adaptor. For a neater, more professional installation, you can cut off the 12 V accessory plug and hardwire this lead into your installation. As accessory plugs can often pop out and work loose, this makes for a better installation.

You will then be left with two leads. One will have a number of phono plugs on the end, one yellow, one red and one white, and a USB plug on the other.

WARNING !!!

Be wary, this USB plug is not intended for connection to a USB port on our Car PC, and you may damage it if you do connect it to a USB port while other devices are connected to it.

Connecting the main monitor lead to the Car PC is straightforward: it consists of USB and VGA connections to our Car PC, and this is illustrated in Figure 5.6. The monitor will have a VGA (monitor) plug on one end, along with a USB type plug and a Mini DIN type connector on the other.

▼   Figure 5.6
    Monitor USB and VGA connections connected (Tim Watson, Photomedia UK)

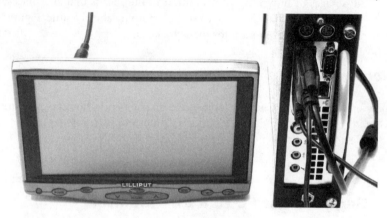

The Mini DIN connector is useful, as it helps us to be able to route the cable without having a monitor on the other end. When we have routed our cable and have our monitor in the right place, we can connect the Mini DIN connector.

The USB plug and lead in fact connects to the USB type socket on our monitor (see Figure 5.8, Connections for sound, composite video and power for details). This is used for composite video. You might use this lead to connect to a DVD player; however, much better quality results will be yielded by using the remaining lead for our Car PC.

Lastly, connect power to the monitor and you are almost ready to rock and roll.

Finally, you need to install the touchscreen drivers. These are supplied on an 8 cm CD ROM. *Do not* under any circumstances insert this into the slot loading drive, it will jam the mechanism. Instead, copy the information on to a memory stick, *or*, burn the drivers to a regular size CD.

Looking at our other options for monitors, mp3car.com sell an *amazing* Xenarc roof mount monitor, shown in Figure 5.9.

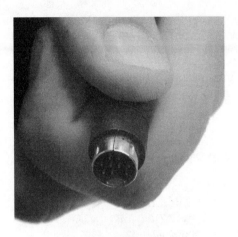

Figure 5.7
Connecting the monitor Mini DIN (Tim Watson, Photomedia UK)

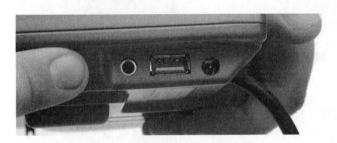

Figure 5.8
Connections for sound, composite video and power (Tim Watson, Photomedia UK)

Figure 5.9
Xenarc 1530YR 15.3" TFT LCD roof mount monitor w/VGA (courtesy Mp3car.com)

Do not be deceived by the picture. The first thing that strikes you about the Xenarc monitor is its sheer size and scale; this thing is *massive*. With a 15.3" widescreen diagonal, you would struggle to fit it in a small vehicle.

One of the great features of this monitor is the infra-red wireless headphones. This enables you to connect your Car PC's audio output to allow passengers to listen to their favorite songs while the driver remains undistracted.

The only thing it is missing is a touchscreen!

We discussed the elegant flip-out monitor supplied by Digital WW and shown in Figure 5.10.

◀ Figure 5.10
*DWW 710 L (courtesy Digital
Worldwide)*

It also comes with a remote control, shown in Figure 5.11. However, if you are investigating controlling your Car PC with total remote (discussed later in the book) then I would strongly advise that you program all of the remote functions into total remote so that you have complete control of your Car PC from your PDA.

There are a large number of connections that need to be made to install this monitor; this is a result of its sophistication and array of features. As you can see in Figure 5.12, there are a LOT of wires to connect.

You will need to connect the red power wire to a permanent 12 V feed, the yellow wire to the ignition switch accessory circuit and the black wire to a good ground.

The next step should be to connect all of the speakers. There is provision for four speakers: front left and right and rear left and right.

If you are going to use this unit's TV function, you will need to connect an aerial that is suitable for in-car use.

Next, you will need to connect an FM aerial for radio reception.

◀ Figure 5.11
*DWW 710L remote control
(courtesy Digital Worldwide)*

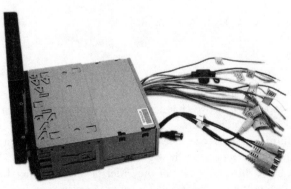

◀ Figure 5.12
*DWW710L wiring (courtesy
Digital Worldwide)*

Be sure when connecting your Car PC to use the VGA inputs as opposed to the composite video input and the TV out on your Car PC. A VGA connection gives a much higher resolution picture.

Finally, you need to connect the USB for the touchscreen and you are away.

In-car audio

For playing music, watching movies, turn by turn directions and interacting with your Car PC, audio is an important function. You will need to connect your Car PC to your car audio system, but as with most things there is a trade-off between convenience and quality.

First of all, let's explore the possibility that for one reason or another, you don't want to actually physically connect your car to your PC's audio.

This might be because:

- You want an "easily removable" system to use in a number of vehicles.
- You have a new car and touching your in-car electronics would invalidate warranty.
- You like your existing stereo but it has no aux inputs.

If that is the case, then you can jump ahead to Chapter 9 and look for the section entitled "Installing the Griffin RocketFM." The Griffin RocketFM is a device that allows you to encode digital audio streams from your Car PC and transmit them as FM radio. This FM radio is beamed to your car stereo where you can select the station as if it were any other pre-set. This is a really great product from Griffin; the only thing is, I hope they release a version 2 with RDS to enable you to display "Car PC FM" on your radio as a preset. That would be awesome. Griffin – if you're reading this!!

If you want to go the hardwires route (which has the edge in terms of sound quality), then you need to consider one of the following scenarios. We will be looking at:

- Wiring your Car PC into your head unit via an "aux" input.
- Wiring your Car PC direct to an amplifier.
- Wiring your Car PC to a surround amplifier.

All use very similar equipment for connection. For each channel you want to connect you will need a male phono jack plug, and for every two channels you will need a 3.5 mm head-phone jack. You will also need screened cable to make the connections; try to keep the lengths as short as possible to minimize interference. The cable between your speakers and amplifier are less susceptible to interference because the signals are at a higher level. The line level input to your amplifier, however, is only ever of the order of one volt, therefore interference from motors, sparks and inductive devices in your car are greatly amplified.

So, for example, for a stereo setup, you will need two phono jacks and a headphone jack, whereas for 5.1 surround you will require six phono jacks and one headphone jack.

Wiring your Car PC into your head unit via an aux input

▼   Figure 5.13
     Connecting your Car PC to your car stereo aux input

Take a look at Figure 5.13; this shows the basic setup in terms of the audio connections between your car head unit, Car PC and speakers.

NOTE

We are only showing the *audio* connections here, all power connections, aerial connections, etc. have been omitted for clarity.

The connections are straightforward. If you want additional control over the sound, you might want to consider inserting a graphics equalizer between the Car PC and the amplifier. Be sure to provide some means of switching the amplifier off with your Car PC. If you are using the Carnetix DC–DC convertor, it has an accessory output you can connect via a relay in order to switch the amp on with your Car PC and ignition.

Wiring your Car PC into an amplifier

Connecting your Car PC to an amplifier differs little compared to connecting your Car PC to a car stereo. The connections are essentially the same; a pair of phono plugs, and the

▼   Figure 5.14
Connecting your Car PC to an amplifier

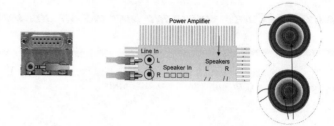

output to a set of speakers is not fundamentally different to a car stereo. However, one thing to bear in mind, is that when routing audio through an amplifier, you have much less control than when you route a Car PC audio through a car stereo.

For a start, an obvious distinction is the fact that with a radio, you have a quickly accessible volume control, whereas with a Car PC the volume control may not always be immediately apparent. For this reason, you might want to consider installing a Griffin Powermate, discussed later in this book.

Wiring your Car PC into a surround amplifier

Wiring in-car surround gives you ultimate control over where the sound is directed, and for movie viewing makes for a much wholler, fuller viewing experience.

How do you go about connecting a surround amplifier to your Car PC? Well, there's more than one way to skin a cat [no one quote me on that to the American Humane Association – I actually really LOVE cats].

First of all, you can adopt an approach whereby all of the analog outputs of your Car PC are connected to the corresponding inputs on an analog surround amplifier. This has the advantage of analog amplifiers without Dolby Digital Decoding being *much* cheaper than more sophisticate models. I have illustrated this approach in Figure 5.15.

The other option is buy a digital amplifier; in this case the connection is much simpler, a single phono-phono coaxial lead.

In-car surround amplifiers are still high-end and "nichey" so it is going to be a while before they become really affordable. There are a couple of good products on the market. First of all, on the web check out: Performance Teknique Amplifier Home Page:

Ⓦ [3] http://www.performance-teknique.com/amplifiers.html

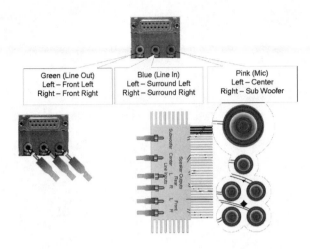

Green (Line Out)
Left – Front Left
Right – Front Right

Blue (Line In)
Left – Surround Left
Right – Surround Right

Pink (Mic)
Left – Center
Right – Sub Woofer

◀ Figure 5.15
*Connecting your Car PC to
an analog 5.1 surround
amplifier*

They produce some really cool high-end kit. Their ICBM-5.1 Amplifier is pretty excellent in every respect, it provides Dolby Digital Decoding on board. This is great news, as rather than having to use 3.5 mm headphone jacks and phono plugs to connect each individual set of stereo channels, we can simply use a single coaxial cable from the phono plug on the VIA board to the amplifier which then performs all of the decoding. What is great about this approach is that your signal can travel further without the degradation inherent in analog connections.

Remember what I said in the last section? You have much less control over the sound with an amplifier than with a head unit. Well, a way around this, is to install something like the Performance Teknique ICBM-5EQ. This gives you a main level control, a "master volume" if you like, a center control, so you can turn up dialog to allow it to be more easily heard over atmospheric effects, and a subwoofer control to allow you to "pump up the bass."

You have also got a fader control, so if the kids want the volume loud in the back, you can let them have it while still retaining some levels of sanity in the front two seats.

Furthermore, you have a balance control so you can move the sound from your side of the car, for example to make sat-nav instructions clearer for you to hear, or to the passenger side.

You also get a subwoofer on/off switch – but then "Smoke on the water" would never sound quite right.

Best of all, and this has to be a number one selling point. The buttons light up blue at night. If you follow this book to the letter your Griffin products and Performance Teknique amp will all color co-ordinate – no clashes.

Chapter 6

On-board diagnostics

For the past 20 years, digital engine management systems have been commonplace in the automobile. Digital engine management systems are bought with the possibility of automatically detecting engine faults, and monitoring variable remotely using electronic sensors and microcontrollers rather than conventional mechanical linkages.

In the early 1990s the automotive industry standardized the method used for diagnosing vehicle errors electronically – this is embodied in the OBD standard.

The systems were originally implemented to help vehicles control their emissions; however, the potential of the system was quickly realized and now OBD is the hub of a car's electrical control system.

The chances are, your vehicle has OBD unless it is an old classic or custom vehicle. On-board diagnostics have been mandatory in the United States since 1988 and are now commonplace throughout the world.

The second generation of the OBD standard, OBD-II, was introduced in 1995. The European Union has obliged manufacturers to incorporate "EOBD" into their vehicles since the year 2000 for petrol engines and 2003 on diesel engines.

For the consumer, this can only be a good thing, as with a few simple tools, you can exploit the functionality of on-board diagnostics using nothing more than your Car PC, some simple software and an OBD interface.

This is great news if you are at all handy with a set of tools, as it means that those "secret manufacturer codes" that can only be sorted out by the dealership when your "check engine light" is illuminated can now be sorted out at home.

When you notice that your vehicle is not quite "tickety boo" you can load up your software and see what is causing your car to throw the dummy out of the pram.

If you're not quite as handy with the tools, it at least allows you to check with your local mechanic with some idea of what is going wrong. If you sound as if you have half an idea about what might be happening, the chances of your getting majorly ripped off decrease significantly.

If you are into modifying vehicles, it means that you can check that what you are doing is actually working.

Looking to the FUTURE ▸ ▸ ▸ ▸ ▸ ▸ ▸ ▸ ▸ ▸ ▸ ▸ ▸ ▸ ▸ ▸ ▸ ▸ ▸

The OBD III standard is in the pipeline. This will enable you to read a vehicle's diagnostic fault codes wirelessly! While this might sound cool, it opens up the scary prospect of the authorities being able to check your vehicle's condition wirelessly, and enforce sanctions if your vehicle is not up to the required standard.

So you have this jumble of wires in your car – how are you going to work out what to do?

The standard ISO 15031-3 says that a 16-pin socket must be fitted near the driver's seat. A lot of earlier cars had this socket fitted in the engine compartment, so if your car is an older model, this is the place to look.

Some common places to find the connector in your vehicle include:

- Below the dashboard
- Next to the fusebox
- Behind a flap in the center console
- Concealed under an ashtray

The connector has 16 pins and looks like Figure 6.1

Figure 6.1
*The (largely standard)
on-board diagnostics connector*

OBD II Pin Assignments

Pin 2	J-1850 bus +
Pin 4	Vehicle Ground
Pin 5	Signal Ground
Pin 6	CAN high (J-2284)
Pin 7	ISO 9141-2 K Output
Pin 10	J- 1850 bus –
Pin 14	CAN low (J-2284)
Pin 15	ISO 9141-2 L Output
Pin 16	Battery Positive

How it all works

A modern car consists of a series of embedded microcontroller systems called ECUs, put simply, "Electronic Control Units." These units talk to each other and share information. For example, the engine management system might have an external air temperature sensor to allow it to calculate air density. The air conditioning system might share this sensor's information for its "External Temperature" display, and similarly, the dashboard system might share this sensor information for its "Frost Warning" light.

By talking to each other over a network, much wiring and circuitry can be saved over a conventional hard-wired system.

It is common to find ECUs in car that control:

- Engine Management
- Anti-Lock Brake Systems
- Electronic Stability Programs
- "Active" Suspension
- Bodywork Functions

As mentioned earlier, these ECUs interact, and to do this they must talk to each other. Electronic devices talk to each other using a hardware system to allow them to communicate, and a protocol – that is to say the information is encoded in a common language.

Many different bus systems and protocols can happily coexist in the same vehicle. Some are more suited to applications due to high speed, for example braking – where rapid calculations mean the difference between life and death; other interfaces and protocols might be less speed critical, and are selected on the grounds of cost or compatibility.

Cracking your car's OBD-II interface

> **WARNING !!!**
>
> Some manufacturers are installing tamperproof connectors in their vehicle to prevent mere mortals such as ourselves from tampering with the OBD-II interface. Some lease and car hire firms already block off certain connectors to prevent access. Playing with these connectors without permission is normally strictly forbidden in the terms of your vehicles lease or warranty and doing so may make them worthless.

With a little help from the folks at "Özen Elektronik" you can build your own OBD interface and in no time at all you will be the "Digital Diagnostic Doctor" able to sort out all of your car's aches and pains.

The brand is certainly a good one, Özen Elektronik are recognized as Official Diagnostic Test Equipment Manufacturer for the Ford Motor Company, so their credentials are sound.

This project is a little more complex than the ones we have tackled before and requires you to have a good working knowledge with electrical components and a soldering iron. If you do not feel confident of your abilities to tackle this project, no sweat, you can buy everything readily assembled from Özen Elektronik who sell a wide range of OBD-II hardware and software products. The interface you are going to build looks like Figure 6.2. Of course how you house it is your business.

◀ Figure 6.2
The mOByDic hardware interface (pre-assembled) (courtesy Özen Elektronik)

This adaptor is *very* versatile indeed; in fact, you can read FIVE different protocols with this adaptor. This is great if you trade up vehicles regularly or if you are considering switching car in the near future, as the chances are this adaptor WILL work on your vehicle.

You will be able to check with your manufacturer which standard your car uses; additionally, you might find information in the vicinity of the OBD-II socket.

For your reference, the protocols that this solution will work with are:

- ISO9141-2
- KWP2000 (ISO142301...4)
- J1850-PWM (SAEJ1850)
- J1850-VPWM (SAEJ1850)
- CAN Bus (ISO-15765-1...4)

What information can I access with this circuit?

The standards currently employed by car manufacturers divide the information you can retrieve from your car's OBD system into "groups." We call these groups "services."

You can access the first four of these services with this circuit; however, changes are constantly being made and you may be able to access more information in the future with an updated chip and/or software from Özen Elektronik.

To give you an overview of the current capabilities of this device:

Service Mode 1

In this mode, you are able to look at the data output from sensors all over your car in "real time."

Service Mode 2

This mode allows you to store data that is stored in your car's control computers. This data is referred to in the industry as "freeze frame" data. This data is primarily stored to enable an assessment to be made of how the engine is performing when running during a normal driving cycle. This allows your service agent to determine whether your car's engine is running within the specified limits set out by the manufacturer. One application of this is to check that the car's emissions are within the prescribed environmental limits.

Service Mode 3

This mode contains the codes that are stored when your "Check Engine Light" illuminates. The codes stored are referred to as "DTCs" in shorthand, which stands for Diagnostic Trouble Codes. These are the ones you want to look at when your car turns ugly.

Service Mode 4

This mode allows you to erase fault codes and stored values. You might want to watch out and be careful though, because in the event of your deleting the fault codes, if the problem persists, your friendly local mechanic will not be able to read the DTCs from your vehicle.

The mOByDic hardware

The circuit we are going to discuss can be built up from parts, or bought as a whole entity from Özen Elektronik. The EPIA SP motherboard used in the Car PC has RS232 support, so we are OK. In the event that this port is occupied, there is a spare header on the motherboard that allows you to connect directly, although this will necessitate a trailing lead out of the back of your Car PC, or a bit of craft metalwork with the rear bezel.

What the mOByDic board does is take the OBD signals, and translates them into a format that can be interpreted through RS232 serial. There are a number of programs that effectively act as "terminals," taking this data and interpreting it.

Although the adaptor supports a vast range of protocols, there is a move in the automotive industry to standardize the CAN protocol.

A microcontroller forms the heart of the mOByDic circuitry. The T89C51 has the ability to communicate with CAN buses thanks to its built-in controller as well as all the other OBD standards.

The chip is running at 32 Mhz. This is because the external 16 Mhz frequency provided by the crystal is multiplied internally. Voltage regulation is provided by that hardy favorite, the 7805 regulator, while RS232 signal conditioning is provided by the MAX232, a pretty vanilla IC, commonplace and easy to get hold of.

In addition to the basic interfacing, the T89C51 drives a pair of status LEDs to give feedback as to the device's status.

The hardware design is such that, whatever interface standard you decide to use, the correct circuitry is connected to the microcontroller to provide level matching.

One possible substitution arises. The PCA82C250 is used to take the signals from the internal CAN controller of the T89C51, and buffers the signals to drive the CAN bus. If you want to use this circuit in a lorry, or a vehicle that uses 24 V, you should substitute the above for a PCA82C251 which is identical in every respect, but is suitable for use with 24 V supplies.

Construction is simple provided that you follow the circuit diagram.

Next we have a complete parts list.

Parts list

1K Resistor (2×)
100R Resistor (2×)
4K7 Resistor (6×)
10K Resistor (9×)
510R Resistor (2×)
1K1 Resistor (1×)
3K9 Resistor (1×)
3K Resistor (1×)
1R Resistor (1×)

10 µF Radial Capacitor 16 V (2×)
27 pF Capacitor (2×)

1 μF Capacitor 16 V (Optional)
100 nF Capacitor (6×)
560 pF Capacitor (2×)

LED High Efficiency Red (1×)
LED High Efficiency Amber (1×)
LED High Efficiency Green (1×)
1N4001 Diode (1×)
1N4148 Diode (2×)
Transistor 2N3906 (2×)
Transistor 2N3904 (3×)

Preprogrammed IC T89C51CC02UA (1×)
(Available from Özen Elektronik Ref 050092-41)

LM339 (DIL 14 Package) (1×)
PCA82C251 or PCA82C250 (DIL 8 Package)(1×)
MAX232 (DIL 16 Package) (1×)
LM7805C (TO220 Package)
LM7808C (TO220 Package)
ZSH560C (TO92 Package) (Optional)

9 Way D-Sub Socket Angled Pins PCB Mount
9 Way D-Sub Plug Angled Pins PCB Mount
2 Way SIL Pinheader
16 Mhz Quartz Crystal 32 pF Parallel Capacitance
28 Way PLCC Socket
Hammond type 1591 B Plastic Case
RS232 Straight Wired Cable (not null modem type)
Adaptor Cable Wired 9 Way D-Sub to OBD Socket
PCB (Available from Özen Elektronik)

NOTE

You will need either the ZSH560C integrated circuit or the 1 μF Capacitor but not both.

If you have the facilities at home, you may want to etch your own printed circuit board.

With Figures 6.3 to 6.5 you should be able to produce a mOByDic PCB. However, save yourself the hassle and mosey on down to

W³ www.ozenelektronik.com

where you can buy the board ready etched.

▼ Figure 6.3
Özen Elektronik mOByDic PCB silkscreen (courtesy Özen Elektronik)

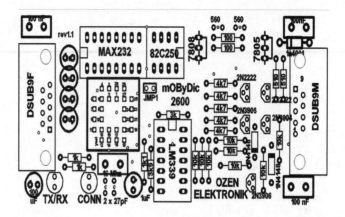

▼ Figure 6.4
Özen Elektronik mOByDic PCB component side layout (courtesy Özen Elektronik)

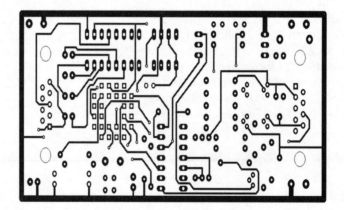

▼ Figure 6.5
Özen Elektronik mOByDic PCB solder side layout (courtesy Özen Elektronik)

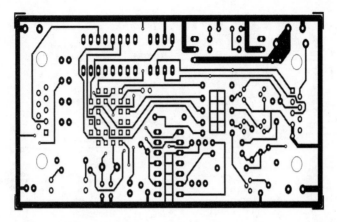

You will need to follow the following circuit diagrams:

▼ Figure 6.6
mOByDic schematic part 1 (courtesy Özen Elektronik)

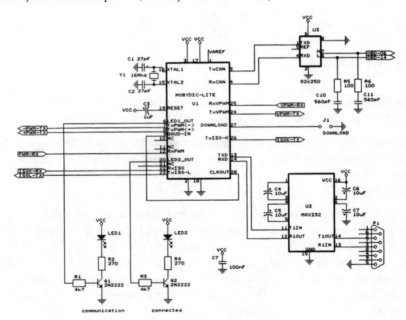

▼ Figure 6.7
mOByDic schematic part 2 (courtesy Özen Elektronik)

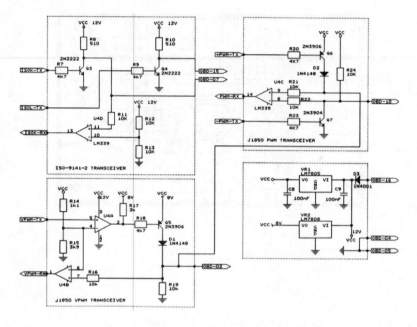

Using the circuit

The interface must be connected to the OBD port by making a suitable lead such as that illustrated in Figure 6.8. You can also buy this cable from Özen Elektronik.

The male nine-way D sub-connector on the interface should be connected to the OBD port, while the female nine-way D sub should be connected to your EPIA SP RS232 port (whether that be the external DB9 or internal header). The cable should be straight through and not null modem (crossed).

The circuit is powered by your vehicle electrics; it should not put any additional strain on the Travla PSU. When you switch your ignition on, the red LED on the mOByDic board will illuminate. Your adaptor will now search for the correct protocol: it has five options, it will check these to see which protocol it can receive and send. The yellow LED will blink in a random fashion while this searching takes place. When it finds the protocol, you will get a green to go light from the green LED. This will not happen if the protocol is not found. In this event, the yellow light will blink on and off.

▼   Figure 6.8
    OBD interface lead (courtesy Özen Elektronik)

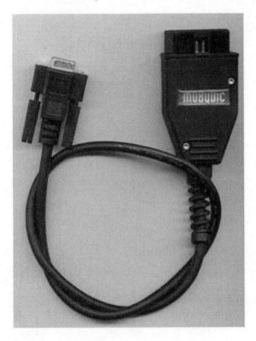

ODB II		DB9
5	————	1
4	————	2
6	————	3
7	————	4
14	————	5
10	————	6
2	————	7
15	————	8
16	————	9

When you get the green light, you can initialize your OBD software and start looking at DTCs.

OBD software

Due to the nature of the mOByDic you will find that there is a wide range of software you can use to read its data; this is because it doesn't use any proprietary technology.

The following links may be of use to you as they all provide software compatible with mOByDic: mOByTester can be downloaded by registering for free at www.ozenelektronik.com. Özen Elektronik provide a wide range of useful and relevant downloads.

Digimoto

   http://www.digimoto.com/

OBD Spy

 http://www.rq-elektronik.de/obd2spy/

Scanmaster

 www.scantool.net/software/scantool.net

Vital Scan DEMO

 http://www.vitalengineering.co.uk/

Comprehensive DTC listing

Here is a listing of all the common DTCs courtesy of Özen Elektronik.

P0100	Mass or Volume Air Flow (MAF) Circuit Malfunction
P0101	Mass or Volume Air Flow (MAF) Circuit Range/Performance Problem
P0102	Mass or Volume Air Flow (MAF) Circuit Low Input
P0103	Mass or Volume Air Flow (MAF) Circuit High Input
P0104	Mass or Volume Air Flow (MAF) Circuit Intermittent
P0105	Manifold Absolute Pressure/Barometric Pressure Circuit Malfunction
P0106	Manifold Absolute Pressure/Barometric Pressure Circuit Range/Performance Problem
P0107	Manifold Absolute Pressure/Barometric Pressure Circuit Low Input
P0108	Manifold Absolute Pressure/Barometric Pressure Circuit High Input
P0109	Manifold Absolute Pressure/Barometric Pressure Circuit Intermittent
P0110	Intake Air Temperature (IAT) Circuit Malfunction
P0111	Intake Air Temperature (IAT) Circuit Range/Performance Problem
P0112	Intake Air Temperature (IAT) Circuit Low Input
P0113	Intake Air Temperature (IAT) Circuit High Input
P0114	Intake Air Temperature (IAT) Circuit Intermittent
P0115	Engine Coolant Temperature (ECT) Circuit Malfunction
P0116	Engine Coolant Temperature (ECT) Circuit Range/Performance Problem
P0117	Engine Coolant Temperature (ECT) Circuit Low Input
P0118	Engine Coolant Temperature (ECT) Circuit High Input
P0119	Engine Coolant Temperature (ECT) Circuit Intermittent
P0120	Throttle Position Sensor (TPS) Circuit Malfunction
P0121	Throttle Position Sensor (TPS) Circuit Range/Performance Problem
P0122	Throttle Position Sensor (TPS) Circuit Low Input
P0123	Throttle Position Sensor (TPS) Circuit High Input

P0124	Throttle Position Sensor (TPS) Circuit Intermittent
P0125	Insufficient Coolant Temperature for Closed Loop Fuel Control
P0126	Insufficient Coolant Temperature for Stable Operation
P0130	O2 Sensor Circuit Malfunction (Bank 1 Sensor 1)
P0131	O2 Sensor Circuit Low Voltage (Bank 1 Sensor 1)
P0132	O2 Sensor Circuit High Voltage (Bank 1 Sensor 1)
P0133	O2 Sensor Circuit Slow Response (Bank 1 Sensor 1)
P0134	O2 Sensor Circuit No Activity Detected (Bank 1 Sensor 1)
P0135	O2 Sensor Heater Circuit Malfunction (Bank 1 Sensor 1)
P0136	O2 Sensor Circuit Malfunction (Bank 1 Sensor 2)
P0137	O2 Sensor Circuit Low Voltage (Bank 1 Sensor 2)
P0138	O2 Sensor Circuit High Voltage (Bank 1 Sensor 2)
P0139	O2 Sensor Circuit Slow Response (Bank 1 Sensor 2)
P0140	O2 Sensor Circuit No Activity Detected (Bank 1 Sensor 2)
P0141	O2 Sensor Heater Circuit Malfunction (Bank 1 Sensor 2)
P0142	O2 Sensor Circuit Malfunction (Bank 1 Sensor 3)
P0143	O2 Sensor Circuit Low Voltage (Bank 1 Sensor 3)
P0144	O2 Sensor Circuit High Voltage (Bank 1 Sensor 3)
P0145	O2 Sensor Circuit Slow Response (Bank 1 Sensor 3)
P0146	O2 Sensor Circuit No Activity Detected (Bank 1 Sensor 3)
P0147	O2 Sensor Heater Circuit Malfunction (Bank 1 Sensor 3)
P0150	O2 Sensor Circuit Malfunction (Bank 2 Sensor 1)
P0151	O2 Sensor Circuit Low Voltage (Bank 2 Sensor 1)
P0152	O2 Sensor Circuit High Voltage (Bank 2 Sensor 1)
P0153	O2 Sensor Circuit Slow Response (Bank 2 Sensor 1)
P0154	O2 Sensor Circuit No Activity Detected (Bank 2 Sensor 1)
P0155	O2 Sensor Heater Circuit Malfunction (Bank 2 Sensor 1)
P0156	O2 Sensor Circuit Malfunction (Bank 2 Sensor 2)
P0157	O2 Sensor Circuit Low Voltage (Bank 2 Sensor 2)
P0158	O2 Sensor Circuit High Voltage (Bank 2 Sensor 2)
P0159	O2 Sensor Circuit Slow Response (Bank 2 Sensor 2)
P0160	O2 Sensor Circuit No Activity Detected (Bank 2 Sensor 2)
P0161	O2 Sensor Heater Circuit Malfunction (Bank 2 Sensor 2)
P0162	O2 Sensor Circuit Malfunction (Bank 2 Sensor 3)
P0163	O2 Sensor Circuit Low Voltage (Bank 2 Sensor 3)
P0164	O2 Sensor Circuit High Voltage (Bank 2 Sensor 3)
P0165	O2 Sensor Circuit Slow Response (Bank 2 Sensor 3)
P0166	O2 Sensor Circuit No Activity Detected (Bank 2 Sensor 3)
P0167	O2 Sensor Heater Circuit Malfunction (Bank 2 Sensor 3)
P0170	Fuel Trim Malfunction (Bank 1)
P0171	System too Lean (Bank 1)
P0172	System too Rich (Bank 1)
P0173	Fuel Trim Malfunction (Bank 2)
P0174	System too Lean (Bank 2)
P0175	System too Rich (Bank 2)
P0176	Fuel Composition Sensor Circuit Malfunction
P0177	Fuel Composition Sensor Circuit Range/Performance

P0178	Fuel Composition Sensor Circuit Low Input
P0179	Fuel Composition Sensor Circuit High Input
P0180	Fuel Temperature Sensor A Circuit Malfunction
P0181	Fuel Temperature Sensor A Circuit Range/Performance
P0182	Fuel Temperature Sensor A Circuit Low Input
P0183	Fuel Temperature Sensor A Circuit High Input
P0184	Fuel Temperature Sensor A Circuit Intermittent
P0185	Fuel Temperature Sensor B Circuit Malfunction
P0186	Fuel Temperature Sensor B Circuit Range/Performance
P0187	Fuel Temperature Sensor B Circuit Low Input
P0188	Fuel Temperature Sensor B Circuit High Input
P0189	Fuel Temperature Sensor B Circuit Intermittent
P0190	Fuel Rail Pressure Sensor Circuit Malfunction
P0191	Fuel Rail Pressure Sensor Circuit Range/Performance
P0192	Fuel Rail Pressure Sensor Circuit Low Input
P0193	Fuel Rail Pressure Sensor Circuit High Input
P0194	Fuel Rail Pressure Sensor Circuit Intermittent
P0195	Engine Oil Temperature Sensor Malfunction
P0196	Engine Oil Temperature Sensor Range/Performance
P0197	Engine Oil Temperature Sensor Low
P0198	Engine Oil Temperature Sensor High
P0199	Engine Oil Temperature Sensor Intermittent
P0200	Injector Circuit Malfunction
P0201	Injector Circuit Malfunction - Cylinder 1
P0202	Injector Circuit Malfunction - Cylinder 2
P0203	Injector Circuit Malfunction - Cylinder 3
P0204	Injector Circuit Malfunction - Cylinder 4
P0205	Injector Circuit Malfunction - Cylinder 5
P0206	Injector Circuit Malfunction - Cylinder 6
P0207	Injector Circuit Malfunction - Cylinder 7
P0208	Injector Circuit Malfunction - Cylinder 8
P0209	Injector Circuit Malfunction - Cylinder 9 or secondary injector #1
P0210	Injector Circuit Malfunction - Cylinder 10 or secondary injector #2
P0211	Injector Circuit Malfunction - Cylinder 11
P0212	Injector Circuit Malfunction - Cylinder 12
P0213	Cold Start Injector 1 Malfunction
P0214	Cold Start Injector 2 Malfunction
P0215	Engine Shutoff Solenoid Malfunction
P0216	Injection Timing Control Circuit Malfunction
P0217	Engine Overtemp Condition
P0218	Transmission Over Temperature Condition
P0219	Engine Overspeed Condition
P0220	Throttle/Petal Position Sensor/Switch B Circuit Malfunction
P0221	Throttle/Petal Position Sensor/Switch B Circuit Range/Performance Problem
P0222	Throttle/Petal Position Sensor/Switch B Circuit Low Input
P0223	Throttle/Petal Position Sensor/Switch B Circuit High Input
P0224	Throttle/Petal Position Sensor/Switch B Circuit Intermittent

P0225	Throttle/Petal Position Sensor/Switch C Circuit Malfunction
P0226	Throttle/Petal Position Sensor/Switch C Circuit Range/Performance Problem
P0227	Throttle/Petal Position Sensor/Switch C Circuit Low Input
P0228	Throttle/Petal Position Sensor/Switch C Circuit High Input
P0229	Throttle/Petal Position Sensor/Switch C Circuit Intermittent
P0230	Fuel Pump Primary Circuit Malfunction
P0231	Fuel Pump Secondary Circuit Low
P0232	Fuel Pump Secondary Circuit High
P0233	Fuel Pump Secondary Circuit Intermittent
P0234	Engine Overboost Condition
P0235	Turbocharger Boost Sensor A Circuit Malfunction
P0236	Turbocharger Boost Sensor A Circuit Range/Performance
P0237	Turbocharger Boost Sensor A Circuit Low
P0238	Turbocharger Boost Sensor A Circuit High
P0239	Turbocharger Boost Sensor B Malfunction
P0240	Turbocharger Boost Sensor B Circuit Range/Performance
P0241	Turbocharger Boost Sensor B Circuit Low
P0242	Turbocharger Boost Sensor B Circuit High
P0243	Turbocharger Wastegate Solenoid A Malfunction
P0244	Turbocharger Wastegate Solenoid A Range/Performance
P0245	Turbocharger Wastegate Solenoid A Low
P0246	Turbocharger Wastegate Solenoid A High
P0247	Turbocharger Wastegate Solenoid B Malfunction
P0248	Turbocharger Wastegate Solenoid B Range/Performance
P0249	Turbocharger Wastegate Solenoid B Low
P0250	Turbocharger Wastegate Solenoid B High
P0251	Injection Pump Fuel Metering Control "A" Malfunction (Cam/Rotor/Injector)
P0252	Injection Pump Fuel Metering Control "A" Range/Performance (Cam/Rotor/Injector)
P0253	Injection Pump Fuel Metering Control "A" Low (Cam/Rotor/Injector)
P0254	Injection Pump Fuel Metering Control "A" High (Cam/Rotor/Injector)
P0255	Injection Pump Fuel Metering Control "A" Intermittent (Cam/Rotor/Injector)
P0256	Injection Pump Fuel Metering Control "B" Malfunction (Cam/Rotor/Injector)
P0257	Injection Pump Fuel Metering Control "B" Range/Performance (Cam/Rotor/Injector)
P0258	Injection Pump Fuel Metering Control "B" Low (Cam/Rotor/Injector)
P0259	Injection Pump Fuel Metering Control "B" High (Cam/Rotor/Injector)
P0260	Injection Pump Fuel Metering Control "B" Intermittent (Cam/Rotor/Injector)
P0261	Cylinder 1 Injector Circuit Low
P0262	Cylinder 1 Injector Circuit High
P0263	Cylinder 1 Contribution/Balance Fault
P0264	Cylinder 2 Injector Circuit Low
P0265	Cylinder 2 Injector Circuit High
P0266	Cylinder 2 Contribution/Balance Fault
P0267	Cylinder 3 Injector Circuit Low
P0268	Cylinder 3 Injector Circuit High
P0269	Cylinder 3 Contribution/Balance Fault
P0270	Cylinder 4 Injector Circuit Low
P0271	Cylinder 4 Injector Circuit High

P0272	Cylinder 4 Contribution/Balance Fault
P0273	Cylinder 5 Injector Circuit Low
P0274	Cylinder 5 Injector Circuit High
P0275	Cylinder 5 Contribution/Balance Fault
P0276	Cylinder 6 Injector Circuit Low
P0277	Cylinder 6 Injector Circuit High
P0278	Cylinder 6 Contribution/Balance Fault
P0279	Cylinder 7 Injector Circuit Low
P0280	Cylinder 7 Injector Circuit High
P0281	Cylinder 7 Contribution/Balance Fault
P0282	Cylinder 8 Injector Circuit Low
P0283	Cylinder 8 Injector Circuit High
P0284	Cylinder 8 Contribution/Balance Fault
P0285	Cylinder 9 Injector Circuit Low
P0286	Cylinder 9 Injector Circuit High
P0287	Cylinder 9 Contribution/Balance Fault
P0288	Cylinder 10 Injector Circuit Low
P0289	Cylinder 10 Injector Circuit High
P0290	Cylinder 10 Contribution/Balance Fault
P0291	Cylinder 11 Injector Circuit Low
P0292	Cylinder 11 Injector Circuit High
P0293	Cylinder 11 Contribution/Balance Fault
P0294	Cylinder 12 Injector Circuit Low
P0295	Cylinder 12 Injector Circuit High
P0296	Cylinder 12 Contribution/Range Fault
P0300	Random/Multiple Cylinder Misfire Detected
P0301	Cylinder 1 Misfire Detected
P0302	Cylinder 2 Misfire Detected
P0303	Cylinder 3 Misfire Detected
P0304	Cylinder 4 Misfire Detected
P0305	Cylinder 5 Misfire Detected
P0306	Cylinder 6 Misfire Detected
P0307	Cylinder 7 Misfire Detected
P0308	Cylinder 8 Misfire Detected
P0309	Cylinder 9 Misfire Detected
P0310	Cylinder 10 Misfire Detected
P0311	Cylinder 11 Misfire Detected
P0312	Cylinder 12 Misfire Detected
P0320	Ignition/Distributor Engine Speed Input Circuit Malfunction
P0321	Ignition/Distributor Engine Speed Input Circuit Range/Performance
P0322	Ignition/Distributor Engine Speed Input Circuit No Signal
P0323	Ignition/Distributor Engine Speed Input Circuit Intermittent
P0325	Knock Sensor 1 Circuit Malfunction (Bank 1 or Single Sensor)
P0326	Knock Sensor 1 Circuit Range/Performance (Bank 1 or Single Sensor)
P0327	Knock Sensor 1 Circuit Low Input (Bank 1 or Single Sensor)
P0328	Knock Sensor 1 Circuit High Input (Bank 1 or Single Sensor)
P0329	Knock Sensor 1 Circuit Intermittent (Bank 1 or Single Sensor)

P0330	Knock Sensor 2 Circuit Malfunction (Bank 2)
P0331	Knock Sensor 2 Circuit Range/Performance (Bank 2)
P0332	Knock Sensor 2 Circuit Low Input (Bank 2)
P0333	Knock Sensor 2 Circuit High Input (Bank 2)
P0334	Knock Sensor 2 Circuit Intermittent (Bank 2)
P0335	Crankshaft Position Sensor A Circuit Malfunction
P0336	Crankshaft Position Sensor A Circuit Range/Performance
P0337	Crankshaft Position Sensor A Circuit Low Input
P0338	Crankshaft Position Sensor A Circuit High Input
P0339	Crankshaft Position Sensor A Circuit Intermittent
P0340	Camshaft Position Sensor Circuit Malfunction
P0341	Camshaft Position Sensor Circuit Range/Performance
P0342	Camshaft Position Sensor Circuit Low Input
P0343	Camshaft Position Sensor Circuit High Input
P0344	Camshaft Position Sensor Circuit Intermittent
P0350	Ignition Coil Primary/Secondary Circuit Malfunction
P0351	Ignition Coil A Primary/Secondary Circuit Malfunction
P0352	Ignition Coil B Primary/Secondary Circuit Malfunction
P0353	Ignition Coil C Primary/Secondary Circuit Malfunction
P0354	Ignition Coil D Primary/Secondary Circuit Malfunction
P0355	Ignition Coil E Primary/Secondary Circuit Malfunction
P0356	Ignition Coil F Primary/Secondary Circuit Malfunction
P0357	Ignition Coil G Primary/Secondary Circuit Malfunction
P0358	Ignition Coil H Primary/Secondary Circuit Malfunction
P0359	Ignition Coil I Primary/Secondary Circuit Malfunction
P0360	Ignition Coil J Primary/Secondary Circuit Malfunction
P0361	Ignition Coil K Primary/Secondary Circuit Malfunction
P0362	Ignition Coil L Primary/Secondary Circuit Malfunction
P0370	Timing Reference High Resolution Signal A Malfunction
P0371	Timing Reference High Resolution Signal A Too Many Pulses
P0372	Timing Reference High Resolution Signal A Too Few Pulses
P0373	Timing Reference High Resolution Signal A Intermittent/Erratic Pulses
P0374	Timing Reference High Resolution Signal A No Pulses
P0375	Timing Reference High Resolution Signal B Malfunction
P0376	Timing Reference High Resolution Signal B Too Many Pulses
P0377	Timing Reference High Resolution Signal B Too Few Pulses
P0378	Timing Reference High Resolution Signal B Intermittent/Erratic Pulses
P0379	Timing Reference High Resolution Signal B No Pulses
P0380	Glow Plug/Heater Circuit "A" Malfunction
P0381	Glow Plug/Heater Indicator Circuit Malfunction
P0382	Glow Plug/Heater Circuit "B" Malfunction
P0385	Crankshaft Position Sensor B Circuit Malfunction
P0386	Crankshaft Position Sensor B Circuit Range/Performance
P0387	Crankshaft Position Sensor B Circuit Low Input
P0388	Crankshaft Position Sensor B Circuit High Input
P0389	Crankshaft Position Sensor B Circuit Intermittent
P0400	Exhaust Gas Recirculation Flow Malfunction

P0401	Exhaust Gas Recirculation Flow Insufficient Detected
P0402	Exhaust Gas Recirculation Flow Excessive Detected
P0403	Exhaust Gas Recirculation Circuit Malfunction
P0404	Exhaust Gas Recirculation Circuit Range/Performance
P0405	Exhaust Gas Recirculation Sensor A Circuit Low
P0406	Exhaust Gas Recirculation Sensor A Circuit High
P0407	Exhaust Gas Recirculation Sensor B Circuit Low
P0408	Exhaust Gas Recirculation Sensor B Circuit High
P0410	Secondary Air Injection System Malfunction
P0411	Secondary Air Injection System Incorrect Flow Detected
P0412	Secondary Air Injection System Switching Valve A Circuit Malfunction
P0413	Secondary Air Injection System Switching Valve A Circuit Open
P0414	Secondary Air Injection System Switching Valve A Circuit Shorted
P0415	Secondary Air Injection System Switching Valve B Circuit Malfunction
P0416	Secondary Air Injection System Switching Valve B Circuit Open
P0417	Secondary Air Injection System Switching Valve B Circuit Shorted
P0418	Secondary Air Injection System Relay "A" Circuit Malfunction
P0419	Secondary Air Injection System Relay "B" Circuit Malfunction
P0420	Catalyst System Efficiency Below Threshold (Bank 1)
P0421	Warm Up Catalyst Efficiency Below Threshold (Bank 1)
P0422	Main Catalyst Efficiency Below Threshold (Bank 1)
P0423	Heated Catalyst Efficiency Below Threshold (Bank 1)
P0424	Heated Catalyst Temperature Below Threshold (Bank 1)
P0430	Catalyst System Efficiency Below Threshold (Bank 2)
P0431	Warm Up Catalyst Efficiency Below Threshold (Bank 2)
P0432	Main Catalyst Efficiency Below Threshold (Bank 2)
P0433	Heated Catalyst Efficiency Below Threshold (Bank 2)
P0434	Heated Catalyst Temperature Below Threshold (Bank 2)
P0440	Evaporative Emission Control System Malfunction
P0441	Evaporative Emission Control System Incorrect Purge Flow
P0442	Evaporative Emission Control System Leak Detected (small leak)
P0443	Evaporative Emission Control System Purge Control Valve Circuit Malfunction
P0444	Evaporative Emission Control System Purge Control Valve Circuit Open
P0445	Evaporative Emission Control System Purge Control Valve Circuit Shorted
P0446	Evaporative Emission Control System Vent Control Circuit Malfunction
P0447	Evaporative Emission Control System Vent Control Circuit Open
P0448	Evaporative Emission Control System Vent Control Circuit Shorted
P0449	Evaporative Emission Control System Vent Valve/Solenoid Circuit Malfunction
P0450	Evaporative Emission Control System Pressure Sensor Malfunction
P0451	Evaporative Emission Control System Pressure Sensor Range/Performance
P0452	Evaporative Emission Control System Pressure Sensor Low Input
P0453	Evaporative Emission Control System Pressure Sensor High Input
P0454	Evaporative Emission Control System Pressure Sensor Intermittent
P0455	Evaporative Emission Control System Leak Detected (gross leak)
P0460	Fuel Level Sensor Circuit Malfunction
P0461	Fuel Level Sensor Circuit Range/Performance
P0462	Fuel Level Sensor Circuit Low Input

P0463	Fuel Level Sensor Circuit High Input
P0464	Fuel Level Sensor Circuit Intermittent
P0465	Purge Flow Sensor Circuit Malfunction
P0466	Purge Flow Sensor Circuit Range/Performance
P0467	Purge Flow Sensor Circuit Low Input
P0468	Purge Flow Sensor Circuit High Input
P0469	Purge Flow Sensor Circuit Intermittent
P0470	Exhaust Pressure Sensor Malfunction
P0471	Exhaust Pressure Sensor Range/Performance
P0472	Exhaust Pressure Sensor Low
P0473	Exhaust Pressure Sensor High
P0474	Exhaust Pressure Sensor Intermittent
P0475	Exhaust Pressure Control Valve Malfunction
P0476	Exhaust Pressure Control Valve Range/Performance
P0477	Exhaust Pressure Control Valve Low
P0478	Exhaust Pressure Control Valve High
P0479	Exhaust Pressure Control Valve Intermittent
P0480	Cooling Fan 1 Control Circuit Malfunction
P0481	Cooling Fan 2 Control Circuit Malfunction
P0482	Cooling Fan 3 Control Circuit Malfunction
P0483	Cooling Fan Rationality Check Malfunction
P0484	Cooling Fan Circuit Over Current
P0485	Cooling Fan Power/Ground Circuit Malfunction
P0500	Vehicle Speed Sensor Malfunction
P0501	Vehicle Speed Sensor Range/Performance
P0502	Vehicle Speed Sensor Low Input
P0503	Vehicle Speed Sensor Intermittent/Erratic/High
P0505	Idle Control System Malfunction
P0506	Idle Control System RPM Lower Than Expected
P0507	Idle Control System RPM Higher Than Expected
P0510	Closed Throttle Position Switch Malfunction
P0520	Engine Oil Pressure Sensor/Switch Circuit Malfunction
P0521	Engine Oil Pressure Sensor/Switch Circuit Range/Performance
P0522	Engine Oil Pressure Sensor/Switch Circuit Low Voltage
P0523	Engine Oil Pressure Sensor/Switch Circuit High Voltage
P0530	A/C Refrigerant Pressure Sensor Circuit Malfunction
P0531	A/C Refrigerant Pressure Sensor Circuit Range/Performance
P0532	A/C Refrigerant Pressure Sensor Circuit Low Input
P0533	A/C Refrigerant Pressure Sensor Circuit High Input
P0534	Air Conditioner Refrigerant Charge Loss
P0550	Power Steering Pressure Sensor Circuit Malfunction
P0551	Power Steering Pressure Sensor Circuit Range/Performance
P0552	Power Steering Pressure Sensor Circuit Low Input
P0553	Power Steering Pressure Sensor Circuit High Input
P0554	Power Steering Pressure Sensor Circuit Intermittent
P0560	System Voltage Malfunction
P0561	System Voltage Unstable

P0562	System Voltage Low
P0563	System Voltage High
P0565	Cruise Control On Signal Malfunction
P0566	Cruise Control Off Signal Malfunction
P0567	Cruise Control Resume Signal Malfunction
P0568	Cruise Control Set Signal Malfunction
P0569	Cruise Control Coast Signal Malfunction
P0570	Cruise Control Accel Signal Malfunction
P0571	Cruise Control/Brake Switch A Circuit Malfunction
P0572	Cruise Control/Brake Switch A Circuit Low
P0573	Cruise Control/Brake Switch A Circuit High
P0574	Cruise Control Related Malfunction
P0575	Cruise Control Related Malfunction
P0576	Cruise Control Related Malfunction
P0577	Cruise Control Related Malfunction
P0578	Cruise Control Related Malfunction
P0579	Cruise Control Related Malfunction
P0580	Cruise Control Related Malfunction
P0600	Serial Communication Link Malfunction
P0601	Internal Control Module Memory Check Sum Error
P0602	Control Module Programming Error
P0603	Internal Control Module Keep Alive Memory (KAM) Error
P0604	Internal Control Module Random Access Memory (RAM) Error
P0605	Internal Control Module Read Only Memory (ROM) Error
P0606	PCM Processor Fault - Watchdog
P0608	Control Module VSS Output "A" Malfunction
P0609	Control Module VSS Output "B" Malfunction
P0620	Generator Control Circuit Malfunction
P0621	Generator Lamp "L" Control Circuit Malfunction
P0622	Generator Field "F" Control Circuit Malfunction
P0650	Malfunction Indicator Lamp (MIL) Control Circuit Malfunction
P0654	Engine RPM Output Circuit Malfunction
P0655	Engine Hot Lamp Output Control Circuit Malfunction
P0656	Fuel Level Output Circuit Malfunction
P0700	Transmission Control System Malfunction
P0701	Transmission Control System Range/Performance
P0702	Transmission Control System Electrical
P0703	Torque Converter/Brake Switch B Circuit Malfunction
P0704	Clutch Switch Input Circuit Malfunction
P0705	Transmission Range Sensor Circuit malfunction (PRNDL Input)
P0706	Transmission Range Sensor Circuit Range/Performance
P0707	Transmission Range Sensor Circuit Low Input
P0708	Transmission Range Sensor Circuit High Input
P0709	Transmission Range Sensor Circuit Intermittent
P0710	Transmission Fluid Temperature Sensor Circuit Malfunction
P0711	Transmission Fluid Temperature Sensor Circuit Range/Performance
P0712	Transmission Fluid Temperature Sensor Circuit Low Input

P0713	Transmission Fluid Temperature Sensor Circuit High Input
P0714	Transmission Fluid Temperature Sensor Circuit Intermittent
P0715	Input/Turbine Speed Sensor Circuit Malfunction
P0716	Input/Turbine Speed Sensor Circuit Range/Performance
P0717	Input/Turbine Speed Sensor Circuit No Signal
P0718	Input/Turbine Speed Sensor Circuit Intermittent
P0719	Torque Converter/Brake Switch B Circuit Low
P0720	Output Speed Sensor Circuit Malfunction
P0721	Output Speed Sensor Range/Performance
P0722	Output Speed Sensor No Signal
P0723	Output Speed Sensor Intermittent
P0724	Torque Converter/Brake Switch B Circuit High
P0725	Engine Speed input Circuit Malfunction
P0726	Engine Speed Input Circuit Range/Performance
P0727	Engine Speed Input Circuit No Signal
P0728	Engine Speed Input Circuit Intermittent
P0730	Incorrect Gear Ratio
P0731	Gear 1 Incorrect ratio
P0732	Gear 2 Incorrect ratio
P0733	Gear 3 Incorrect ratio
P0734	Gear 4 Incorrect ratio
P0735	Gear 5 Incorrect ratio
P0736	Reverse incorrect gear ratio
P0740	Torque Converter Clutch Circuit Malfunction
P0741	Torque Converter Clutch Circuit Performance or Stuck Off
P0742	Torque Converter Clutch Circuit Stuck On
P0743	Torque Converter Clutch Circuit Electrical
P0744	Torque Converter Clutch Circuit Intermittent
P0745	Pressure Control Solenoid Malfunction
P0746	Pressure Control Solenoid Performance or Stuck Off
P0747	Pressure Control Solenoid Stuck On
P0748	Pressure Control Solenoid Electrical
P0749	Pressure Control Solenoid Intermittent
P0750	Shift Solenoid A Malfunction
P0751	Shift Solenoid A Performance or Stuck Off
P0752	Shift Solenoid A Stuck On
P0753	Shift Solenoid A Electrical
P0754	Shift Solenoid A Intermittent
P0755	Shift Solenoid B Malfunction
P0756	Shift Solenoid B Performance or Stuck Off
P0757	Shift Solenoid B Stuck On
P0758	Shift Solenoid B Electrical
P0759	Shift Solenoid B Intermittent
P0760	Shift Solenoid C Malfunction
P0761	Shift Solenoid C Performance or Stuck Off
P0762	Shift Solenoid C Stuck On
P0763	Shift Solenoid C Electrical

P0764	Shift Solenoid C Intermittent
P0765	Shift Solenoid D Malfunction
P0766	Shift Solenoid D Performance or Stuck Off
P0767	Shift Solenoid D Stuck On
P0768	Shift Solenoid D Electrical
P0769	Shift Solenoid D Intermittent
P0770	Shift Solenoid E Malfunction
P0771	Shift Solenoid E Performance or Stuck Off
P0772	Shift Solenoid E Stuck On
P0773	Shift Solenoid E Electrical
P0774	Shift Solenoid E Intermittent
P0780	Shift Malfunction
P0781	1-2 Shift Malfunction
P0782	2-3 Shift Malfunction
P0783	3-4 Shift Malfunction
P0784	4-5 Shift Malfunction
P0785	Shift/Timing Solenoid Malfunction
P0786	Shift/Timing Solenoid Range/Performance
P0787	Shift/Timing Solenoid Low
P0788	Shift/Timing Solenoid High
P0789	Shift/Timing Solenoid Intermittent
P0790	Normal/Performance Switch Circuit Malfunction
P0801	Reverse Inhibit Control Circuit Malfunction
P0803	1-4 Upshift (Skip Shift) Solenoid Control Circuit Malfunction
P0804	1-4 Upshift (Skip Shift) Lamp Control Circuit Malfunction
P1100	Fuel Pump - #1 Relay Open-Circuit
P1101	Fuel Pump - #1 Relay Short-Circuit
P1102	Fuel Pump - #2 Relay Open-Circuit
P1103	Fuel Pump - #2 Relay Short-Circuit
P1106	Barometric (BARO) sensor performance
P1107	Barometric (BARO) sensor circuit intermittent low voltage
P1108	Barometric (BARO) sensor circuit intermittent high voltage
P1111	Intake Air Temperature (IAT) sensor intermittent high voltage
P1112	Intake Air Temperature (IAT) sensor intermittent low voltage
P1114	Engine Coolant Temperature (ECT) sensor circuit intermittent low voltage
P1115	Engine Coolant Temperature (ECT) sensor circuit intermittent high voltage
P1121	Throttle Position (TP) sensor inconsistent with MAF Sensor high voltage
P1122	Throttle Position (TP) sensor inconsistent with MAF Sensor low voltage
P1133	Heated Oxygen Sensor (HO2S) insufficient switching bank 1 sensor 1 (Rear Bank)
P1134	Heated Oxygen Sensor (HO2S) transition time ratio bank 1 sensor 1 (Rear Bank)
P1153	Heated Oxygen Sensor (HO2S) insufficient switching bank 2 sensor 1 (Front Bank)
P1154	Heated Oxygen Sensor (HO2S) transition time ratio bank 2 sensor 1 (Front Bank)
P1189	Engine Oil Pressure Switch Circuit
P1231	Fuel pump Sekund"rkreis
P1258	Engine Metal Over Temperature Protection
P1301	Misfire - Emissions Increase
P1302	Misfire - Catalyst Damage

P1311	Rough Road Sensor Open-Circuit
P1312	Rough Road Sensor Short-Circuit
P1313	Rough Road Sensor Performance
P1320	Ignition Control (IC) Module 4x Reference Circuit Intermittent No Pulses
P1323	Ignition Control (IC) Module 24x Reference Circuit low frequency
P1350	Ignition Control System
P1370	Ignition Control (IC) Module 4x Reference too many pulses
P1371	Ignition Control (IC) Module 4x Reference too few pulses
P1375	Ignition Control (IC) Module 24x Reference High Voltage
P1376	Ignition Ground Circuit
P1377	Ignition Control (IC) Module Cam Pulse to 4x Reference Pulse Comparison
P1380	EBTCM DTC Detected-Rough Data Unusable
P1381	Misfire Detected-No EBTCM/PCM Serial Data
P1400	Coolant Fan Relay Short-Circuit
P1401	Coolant Fan Relay Open-Circuit
P1404	Exhaust Gas Recirculation (EGR) Valve Pintle Stuck Open
P1405	Exhaust Gas Recirculation (EGR) Sensor
P1406	Exhaust Gas Recirculation (EGR) Temp Sensor Above High Limit
P1407	Exhaust Gas Recirculation (EGR) Temp Sensor Below Low Limit
P1410	Wastegate Solenoid Short-Circuit
P1411	Wastegate Solenoid Open-Circuit
P1412	Wastegate Boost Control Function
P1420	Battery Voltage - High
P1421	Battery Voltage - Low
P1425	Catalyst System Failing - Right-Hand Bank
P1435	Catalyst System Failing - Left-Hand Bank
P1441	Evaporative System Flow During Non-Purge
P1460	A/C Control Relay Open-Circuit
P1461	A/C Control Relay Short-Circuit
P1470	Coolant Re-circ Relay Open-Circuit
P1471	Coolant Re-circ Relay Short-Circuit
P1483	Engine Cooling System Performance
P1508	Idle Air Control (IAC) System -Low RPM
P1509	Idle Air Control (IAC) System -High RPM
P1520	Intake Manifold Runner Control (IMRC) circuit malfunction
P1524	Throttle Position (TP) Sensor Learned Closed Throttle Angle Degrees Out-Of-Range
P1527	Trans Range / Pressure Switch Comparison
P1530	AC Compressor relays Prim"rkreis
P1540	PWM - AC Pressure Signal
P1554	Cruise Engaged Circuit High Voltage
P1560	Cruise Control System-Transaxle Not in Drive
P1564	Cruise Control System-Vehicle Acceleration too high
P1566	Cruise Control System-Engine RPM Too High
P1567	Cruise Control - ABCS Active
P1570	Cruise Control System - Traction Control Active
P1571	Traction Control System PWM Circuit No Frequency
P1574	EBTCM System - Stop Lamp Switch Circuit High Voltage

P1575	Extended Travel Brake Switch Circuit High Voltage
P1579	Park/Neutral to Drive/Reverse At high Throttle Angle
P1599	Engine Stall or Near Stall Detected
P1602	Loss of EBTCM Serial Data
P1603	Loss of SDM Serial Data
P1604	Loss of IPC Serial Data
P1605	Loss of HVACC Serial Data
P1610	Loss of PZM Serial Data
P1611	Loss of CVRTD Serial Data
P1612	Wegfahrsperre: Challenge request not received
P1617	Engine Oil Level Switch Circuit
P1621	PCM Memory Performance
P1626	Theft Deterrent System - Fuel Enable Circuit
P1630	Theft Deterrent - PCM in Learn Mode
P1631	Theft Deterrent - Password Incorrect
P1632	Theft Deterrent - Fuel Disabled
P1633	Ignition Supplement Power Circuit Low Voltage
P1634	Ignition 1 Power Circuit Low Voltage
P1640	Driver 1 - Input High Voltage
P1641	Malfunction Indicator Lamp (MIL) Control Circuit
P1642	Vehicle Speed Output Circuit
P1644	Delivered Torque Output Circuit
P1645	EVAP Solenoid Output Circuit
P1646	EVAP Vent Valve Output Circuit
P1650	Driver 2 - Input High Voltage
P1652	Lift/Dive Circuit
P1654	Cruise Disable Output Circuit
P1660	Cooling Fan Control Circuits
P1813	Transmission signal
P9999	Undefined fault

Chapter 7

In-car speech recognition

What's the point?

Speaking to your car computer system is undoubtedly cool, David Hasselhoff was doing it as Michael Knight in the eighties but now science fiction can truly be reality. Once your car is equipped with a PC, you can take it one step further to the ultimate level of geek chic and enable your car to receive spoken commands.

The reasoning behind this is very sound. While driving a car, there are a lot of demands placed on you. You really need your attention fully fixed on the road, not fumbling around with knobs and buttons in order to drive safely. If you can remove some of the burden from the driver's hands – by allowing them to perform non-safety critical functions with the voice rather than the hands, then the driver can be allowed to concentrate their attention on the road rather than the radio.

So stick a car PC in an old Trans Am, go to a dodgy hair stylist and you're away.

OUT OF THE BOX

Some of the things your Car PC might say to you shortly after you are through with this chapter and have installed voice recognition:

"Please Michael, I'm the Knight Industries 2000, not a tomato on wheels!"

"Going through walls isn't my favorite pastime, but it sure beats socializing with a donkey!"

"You stick me in storage for over a decade, then sell off my parts like I'm inventory for Manny, Moe and Jack!"

In this chapter, we are going to be implementing a combined solution of hardware and software that makes for accurate speech recognition. Software is improving all the time, however there are still some issues and the technology isn't perfect. With our Car PC applications we will only be using speech recognition to control non-safety critical applications, so in the worst case scenario, you will end up listening to "Country and Western" rather than "Heavy Metal."

First of all, let's briefly discuss speech recognition technology.

Styles of speech recognition

Speech recognition has developed enormously to the extent that naturally spoken speech can be recognized with great accuracy with what would now be considered a fairly modest computer. Depending on the application, and whether the spoken voice being recognized is of one user or a variety of users, there are a number of different "styles" of speech recognition, each being appropriate for different applications.

The following descriptions come from my other great book *50 Awesome Auto Projects for the Evil Genius*

Isolated

In this style of speech recognition, only individual words are recognized. This is the main type of recognition you will work with in the hardware based speech recognition project in this chapter.

Connected

In connected speech recognition, a number of words can be recognized in short succession. Using the longer word length, it is possible to experiment with this type of speech recognition with the hardware-based project in this chapter.

Continuous

Continuous speech recognition systems can recognize naturally spoken, continuous unbroken speech. Some PC-based software applications are approaching good reliability with continuously spoken speech; however, further development is still required.

How many users?

Speaker dependent

A speaker dependent system can only recognize a user for which it is has been trained. Examples include dictating a letter to your word processor.

Speaker independent

A speaker independent system can work with any user. An example would be calling your bank where the voicemail system recognizes the number you say.

Does it really work?

Yes! It does. There are *lots* of people with Car PC setups who complain of poor speech recognition accuracy while driving. The problem is that the microphones they are using are wholly unsuited to the job. Your basic microphone is unidirectional, which means that it receives sound from all around. Clearly this is not ideal. Slightly more advanced microphones have one mike pointing towards the speaker, and one pointing away, with a simple electrical circuit that subtracts the sound from the microphone pointing away from the one pointing towards the driver. Even so, the voice recognition software is decoding the sound from the whole of that direction, which still leaves a lot of noise for the software to have to filter. The Andrea Electronics DA-350 array (see Figure 7.1) uses a wholly different technology.

By combining a linear array of microphones with an advanced digital signal processor, the array filters out extraneous noise and pinpoints the driver as the source.

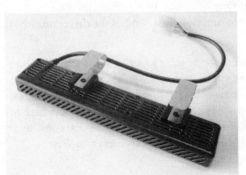

Figure 7.1
The DA-350 linear array microphone

This microphone was tested with a Clarion Auto PC and received their approval.

You have to bear in mind that the Clarion is a far less sophisticated beast than your Car PC. We discussed its specs earlier in Chapter 1. With your Car PC's improved voice recognition abilities, accuracy will improve significantly.

TEST Situation

A standard directional microphone commonly used in the automotive industry was tested against the Andrea Electronics DA-350. A Clarion AutoPC was used for speech recognition, with 28 single word spoken commands used as the basis for recognition accuracy. This test rig was used in three different scenarios.

1. The first scenario was voice recognition while driving with one window partially open and the radio on.
2. The second scenario was voice recognition while driving with one window FULLY open and the radio loud.
3. The third scenario was voice recognition performed with BOTH windows open and the radio loud.

The test results are reprinted in Table 7.1.

This array is so effective, that it is used by the police in a system that allows officers to read out number plates of vehicles they are in pursuit of, and have information read back to them from a database. When you imagine the amount of noises in a police car: siren wailing, engine running, radio blaring, road noise, not to mention the sound of the driver munching noisily on his burger.

The results clearly show that the linear array microphone is far superior to the standard directional microphone, as used in many automotive applications and Car PC setups. By retaining your existing software and upgrading to a linear array microphone, the performance of a Car PC setup can be greatly enhanced. This is especially telling in the third scenario where the radio is on loud and both windows are wound down. We can see that the DA-350 consistently recognizes words, whereas the standard directional microphone fails every time.

Table 7.1

	Test #1				Test #2				Test #3			
Microphone	Competitor		Andrea Array		Competitor		Andrea Array		Competitor		Andrea Array	
Description	Automobile Radio On, Window Partial		Automobile Radio On, Window Partial		Automobile Radio Loud, Window Down		Automobile Radio Loud. Window Down		Automobile Radio Loud, Both Windows Down		Automobile Radio Loud, Both Windows Down	
Radio	Score	# Tries	Score	# Tries	Score	# Tries	Score	# Tries	Score	# Tries	Score	# Tries
Nomad	59	1	71	1	66	3	57	1	no recognition		68	1
Start	55	1	62	1	50	2	58	1	no recognition		54	1
Mute	61	1	63	1	58	3	66	1	no recognition		69	1
Volume	61	1	70	1	56	4	60	1	no recognition		68	1
Help	68	1	59	1	50	2	67	1	no recognition		63	1
Next	60	1	55	1	61	1	63	1	no recognition		66	1
Previous	71	1	71	1	67	5	70	1	no recognition		70	1
Radio	70	1	70	1	71	2	71	1	no recognition		74	1
Disc Player	79	1	79	1	73	1	79	1	no recognition		70	1
Navigate	70	1	72	1	62	2	71	1	no recognition		66	1
Address Book	66	2	69	1	51	2	60	1	no recognition		52	1
Setup	52	1	70	1	53	2	68	4	no recognition		81	1
Audio	68	1	83	1	65	1	69	1	no recognition		65	1
AM	66	1	73	1	57	6	69	1	no recognition		64	3
FM	57	1	62	1	69	1	52	2	no recognition		72	1
Preset	69	1	71	1	63	1	66	1	no recognition		52	1
Stop	60	1	57	1	58	2	59	1	no recognition		75	1
Review	75	2	74	1	76	3	79	1	no recognition		62	1
Zero	72	1	69	1	58	1	74	1	no recognition		70	1
One	65	1	73	1	65	1	72	1	no recognition		64	1
Two	63	1	61	1	62	1	69	1	no recognition		56	1
Three	53	1	65	1	67	4	73	1	no recognition		60	1
Four	59	1	60	1	55	1	57	1	no recognition		73	1
Five	72	1	75	1	70	1	74	1	no recognition		80	1
Six	79	1	74	1	63	1	83	1	no recognition		64	1
Seven	54	1	51	1	54	1	72	1	no recognition		70	1
Eight	58	3	60	2	50	3	72	1	no recognition		56	1
Nine	72	1	74	1	54	1	65	3	no recognition		56	1
	64.8		67.6		60.9		67.7		0		65.7	
	87.50%		96.60%		49.10%		82.40%		0.00%		93.30%	

How does the technology work?

If we refer to the diagram in Figure 7.2, we can see the following taking place. We can see that there is a driver producing an audio signal. Either side of the driver there are noisy wave forms. Voice recognition in the automotive environment is hard to achieve with conventional technology; noise comes from many sources – unless you are in the middle of the wilderness with the car parked, engine off, windows wound up, radio muted, and kids gagged.

▼   Figure 7.2
     How the DSP works to filter extraneous signals (courtesy Andrea Electronics)

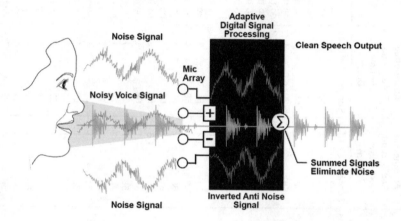

The problem is that all of the extraneous noise present in a car provides a noisy signal that the computer finds hard to interpret. The driver's voice becomes lost in a sea of electrical fuzz.

Rather than speaking to a single microphone (as would be the case with a standard directional microphone), the driver speaks to a linear array of microphones. This is simply a posh way of saying a number of microphones in a line.

Now for the clever bit! The information from this "array" of microphones is fed into a digital signal processor. These are sophisticated chips that have the ability to process a large amount of voice information. This results in a very directional microphone. The polar plot in Figure 7.3 indicates the best positioning for optimal speech recognition response. Figure 7.4 illustrates the response of the microphone at various angles as shown on the polar plot.

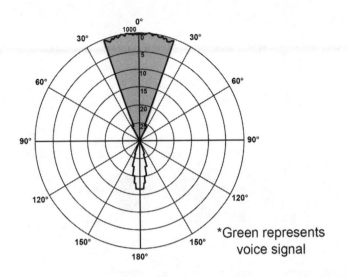

◄ Figure 7.3
*Polar plot of
DA-350 response
(courtesy Andrea
Electronics)*

*Green represents
voice signal

▼ Figure 7.4
Response of the microphone at various angles (courtesy Andrea Electronics)

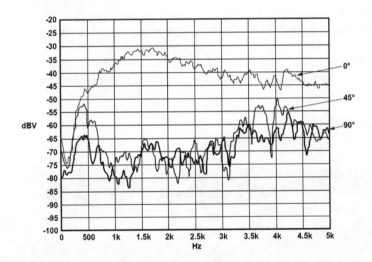

Hardware installation

Hardware Installation of the DA-350 is relatively easy "out of the box."

Physically installing the unit is very easy. Figure 7.5 gives the physical dimensions of the microphone, which should help immensely with working out the space in your vehicle.

135

▼ Figure 7.5
The physical dimensions of the DA-350 (courtesy Andrea Electronics)

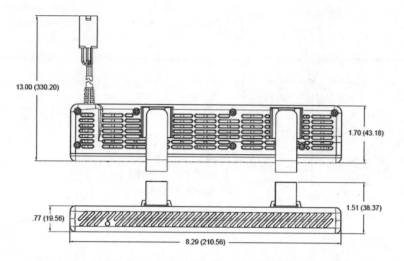

The array is a nice size that will easily clip onto your sun visor (see Figure 7.6); however, I feel that an integrated installation looks really good if you have the skill to make it happen. The metal clips on the DA-350 can easily be removed for direct mounting to the roof, and the cable can easily be run down an A-pillar, although you should realize that you will probably invalidate your warranty on the microphone if you modify the casing.

▼ Figure 7.6
The D-350 clipped to your sun visor

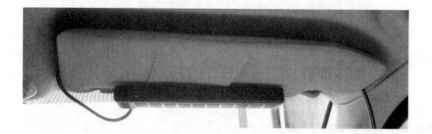

The leads supplied with the DA-350 have a connector in the middle, which makes routing cables a lot easier. When routing the cables, push the narrow middle connectors through any small openings and join them at the middle. This is much better than enlarging holes to accept the large cigarette lighter plug and audio jack. When you have worked out the positioning of your cables and routed them, join them by connecting the connector as shown in Figure 7.7. Make sure you get the orientation of the connector right, it is hard not to!

▼ Figure 7.7
Joining the DA-350 connector

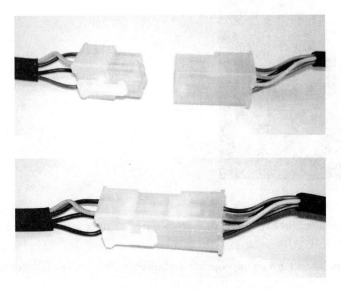

There are two principle connections, power and audio. These are shown in Figure 7.8. The power is provided by a large cigarette lighter style plug, and the audio by a standard 3.5 mm jack.

▼ Figure 7.8
Power and audio connections

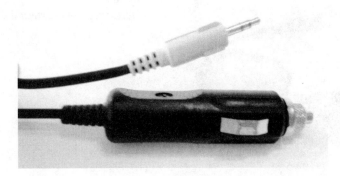

The cigarette lighter plug just fits into your standard 12 V accessory socket as shown in Figure 7.9.

Figure 7.9
Power connected to the DA-350

Alternatively, you might like to use an in-line cigarette lighter socket to connect the microphone up out of the way.

For those who haven't got an appropriate audio connector, Andrea Electronics also supply an external USB interface, as shown in Figure 7.10.

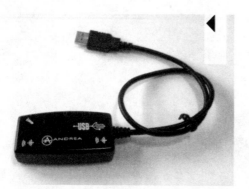

Figure 7.10
Andrea electronics external interface

Once all these connections are made, the hardware installation of your microphone is complete.

The main steps to be followed for successful implementation of a voice control system are:

Hardware installation (covered above)

Software installation

Setting up audio and microphone properties

Creating a model of your voice

Setting up audio and microphone properties

You need to set up your preferences in the "Recognition Profile Settings" of Windows XP. To do this, go to the control panel, and click on "Speech."

This will bring up the window in Figure 7.11. First we want to click on "Settings." This brings up the Window shown in 7.12. Now make the following changes.

◀ Figure 7.11
Speech properties

Pronunciation Sensitivity

You need to move this slider from its default center position to the right.

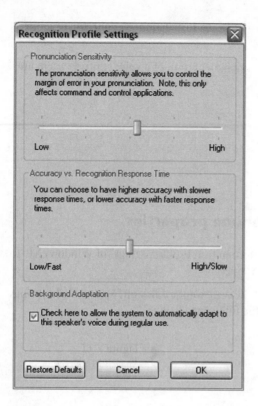

Figure 7.12
Recognition profile settings

Accuracy vs Recognition Response Time

You need to move this slider from its default center position to the right.

Background Adaptation

You will see this box is ticked by default. I highly recommend that you uncheck it. This will prevent your PC trying to learn from its background environment. When you think about the extraneous noise sources in the car, learning from its environment would be undesirable.

Now click OK. This takes you back to Figure 7.11 "Speech Properties."

Language

This allows you to select the desired speech recognition engine, and the language you will be using for recognition.

Recognition Profiles

This allows you to configure different profiles for different people and noise environments. One thing that has proved successful is creating different profiles for stationary and moving. You can use the stationary profile while parked in a low noise environment, and the moving profile for voice recognition on the road.

Audio Input

This allows you to select the input that your microphone is connected to. If you are using the DA-350 array with the external USB option, you will need to change the setting in this window.

Configure Microphone

Clicking on this box brings up a wizard that enables you to set the level of your microphone. Work through this before attempting speech recognition.

Voice recognition software

IBM Via Voice Advanced from Scansoft provides a great set of command and control functionality for your Car PC. The software is well worth the extra couple of bucks over the "Personal" and "Standard" versions as it offers much better "Command and Control" options, which after all is what you will be using the software for primarily.

Getting to grips with voice control software can take a bit of time. The manual for Via Voice Advanced is a 148-page tome, due to the fact that the software is very sophisticated and has a lot of functionality.

It would be tedious going into every detail of the software here when the user manual is freely downloadable from

 www.scansoft.com

but hopefully I can give you an overview of the software, and guidance as to what features are relevant in a car computing context.

Voice analysis

Via Voice allows you to start any program that is on the desktop or in the Windows "Programs" menu by saying "Start Program…" where "…" is the name of the program.

The other great thing is that the software allows you to minimize, restore and all the other things that you would normally do with a mouse to control programs on your desktop, without taking your hands off the wheel or taking your attention off the road.

Say, for example that you have your present position shown on screen in Map Point, you might want to say "Minimize MapPoint" and then "Start Program Windows Media Player."

If you find you have difficulty in getting the Car PC to recognize your natural commands, you can give the Car PC an "attention word." The Car PC listens for this word in order to know when it is being spoken to, the same way as in natural conversation you might say "Jane, would you mind if…" the word "Jane" indicates to Jane that you are speaking to her. Similarly, you could set up a boring "attention word" like "Computer," or something dumb, or even better you could give your Car PC a name. Like Jane or KITT or Priscilla or Quentin or Elvis or Hank or Judy…

You can set the attention word by going to the "Dictation Tab" in the Via Voice software. Alternatively, if you want to give your Car PC a name, you can set multiple attention words from the "Command Sets" tab in the Via Voice Options.

Another great function of Via Voice Advanced that you might like to explore is "Voice Mouse." This allows you to move the cursor on the screen by simply saying "Move…" followed by the direction to move in and the distance. For hands-free navigation, this feature is a real boon.

There are also a number of programs that have been written by Car PC enthusiasts based upon the Microsoft SAPI. You might want to search the mp3car.com forums for "Map Monkey" and "Navi Voice," two very useful applications.

Chapter 8

Killer software applications for your Car PC

Music, GPS and Multimedia in cars are all pretty commonplace now, but there are some features that it would be great to have in your car that you just can't get off the shelf.

This chapter deals with readily available software, some of it free or shareware, that enables you to get the most out of your Car PC. You have made a real investment building your Car PC; some of the tools on the pages that follow, will, with regular use, save you significant amounts of money on your vehicle.

Lap timer

 http://www.gregorybraun.com/LapTimer.html

Lap Timer is a freeware piece of software that was designed to allow users to time laps on any scale slot-car track. The slot car's tracks are timed to 1/1000th second accuracy. So what is there to stop you from timing your own car as you race it round a circuit?

Rather than mounting photocells in the track to detect cars passing overhead, fit one on your dashboard, and every time your finger covers it as you complete a lap, you time that lap. Simple, eh?

Lap Timer is a free download and comes with constructional details for a number of cables that allow it to interface to a Joystick Port, Serial Port, Printer Port or Keyboard Port. The wiring diagrams are easy to follow and simple to construct. When entering track parameters, you will need to take into account the scale, and alter things about a little bit. You can trigger your lap timer manually on the dashboard.

The software allows you to time up to four lanes, although in practice, you would only really want to use one lane.

The program then allows you to tabulate the information into "best lap time," and then break the remainder down into individual lap times.

Automotive Wolf

 http://www.lonewolf-software.com/

Figure 8.1
Automotive Wolf box shot (courtesy Lone Wolf Software)

Automotive Wolf is a sophisticated piece of "Car Management Software" that allows you to keep detailed maintenance and repair records for your vehicle. It also prompts you when essential service and repairs need to be done with its built-in maintenance reminders – or you even have the option to set your own. The software enables you to set whether you hear the maintenance alarms when you launch the program, or boot your Car PC. This is a really useful feature, as it means that every time you jump into your car and switch on the ignition, you get essential status information as to the condition of your car.

The great thing is, when your car does need any routine or preventative maintenance, Automotive Wolf will generate an inventory of the parts required.

The software accurately predicts when car maintenance and inspections are due. Because vehicles and driving styles vary so much the software comes with configurable maintenance/inspection requirements.

In addition to this, you learn more about your vehicle every time you use the software with the Car Care Tips section.

If you get stuck by the roadside, and you are trying to understand why something broke down in order to repair it, you can use the Automotive Systems Reference to help you.

In addition to this, the program will also help you with automotive diagnosis and troubleshooting! A computer is no substitute for a real mechanic, but certain so-called "expert

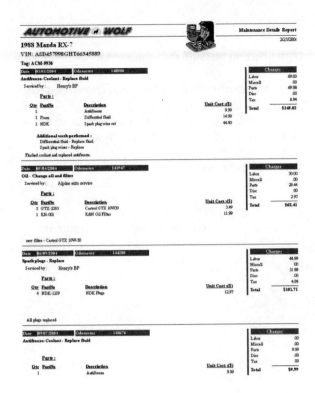

Figure 8.2
*Example of Automotive
Wolf maintenance
report (courtesy Lone
Wolf Software)*

systems" can help you sort out a problem by asking questions and using a funneling technique to narrow down the possibilities until you find the correct answer.

It's bit like a Windows Troubleshooter. But then again – how many people have ever managed to do anything useful with a Windows Troubleshooter?

By using Detailed Cost Analysis you can calculate useful figures like cost/mile, cost/year, which give you a better idea of what it is costing to run your vehicle. It will also give you information on the investment value of your vehicle.

By looking at the whole life of your vehicle, you can see whether it was a sound investment or not. A cheaper vehicle might cost less to begin with, but with this software, you can examine the accumulated costs and operating costs of your vehicle.

Do you drive your car heavy? With spiraling fuel costs, a program that tracks fuel usage and fuel economy might be very beneficial in helping you to develop better driving habits.

What is particularly nice is that this program displays timely information in the form of gauges, allowing you to view at a glance the condition of various vehicle components.

The software includes the capability for:

- Three digital LED gauges indicate the current state of three important filters on each vehicle.
- Six functional analog gauges display the status of any maintenance requirement you select or create!
- Vehicle status gauge that monitors the overall condition of each vehicle with respect to the status of all enabled maintenance requirements configured for each vehicle.

See Figure 8.3 for an example of Automotive Wolf Gauges.

Figure 8.3
Automotive Wolf Gauges (courtesy Lone Wolf Software)

MileMate

 http://www.milemate.com/

MileMate allows you to keep a handle on your car's expenses, by letting you keep a record of oil changes, tax, registration expenses, insurance, letting you monitor where your money is going, and helping you to keep on top of car maintenance.

MileMate allows you to display information about the service status of various pieces of car equipment as a "dashboard" display on your Car PC. Figure 8.5 shows an example of a MileMate dashboard.

You can get MileMate in three versions, which the company calls "trim levels" – a kind of cutesy touch!

MileMate CLe, the classic edition, comes with modules that allow you to perform trip tracking and fuel consumption monitoring. The program comes with helpful hints on vehicle maintenance, how to service your car at home and also how to use MileMate.

You can monitor an unlimited amount of vehicles. MileMate puts each vehicle in a "garage" so the automotive theme is continued throughout the software.

MileMate DXe is the deluxe edition which is a happy compromise between the "CLe" and "MSe" editions. This package provides more vehicle maintenance functions as well as more

▼ Figure 8.4
Automotive Wolf screen shots (courtesy Lone Wolf Software)

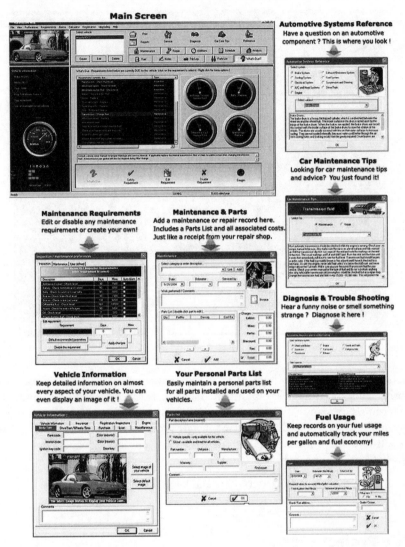

vehicle performance monitoring options. You can set up service reminders for up to four individual items on your vehicle, and a better, improved dashboard compared to the CLe variant. You can set the reminders to work based on either time, distance traveled, service history or the vehicle's age.

▼ Figure 8.5
    MileMate MSe dashboard (courtesy Blue Sunset Software)

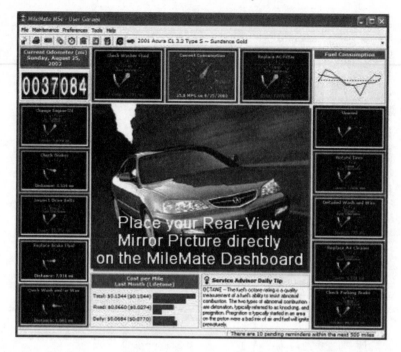

MileMate MSe, the master edition of MileMate, is the crème de la crème of dashboard monitoring software. It has even more functions and features than the CLe and DXe variants. You can set as many different service reminders as you want with up to 12 reminders showing on the MileMate dashboard. The software provides in-depth cost analysis functions and the software will produce a full parts inventory for your local service depot, telling them exactly what work you need done on the vehicle.

You can also use the software in both miles and kilometers.

Vehicle Project Planner

 www.peedle.com

When you are about to start a big project on your vehicle, it makes sense to plan ahead. Furthermore, when you are in the middle of the project, it makes sense to look at what you have spent on the project so that the costs don't run away with themselves. [Author's note: If my father could hear me now, he would be shouting "hypocrite" at me and telling me to practice what I preach!"]

Vehicle Project Planner does just that! It allows you to manage expenditure on major vehicle projects, and look back at how much you have spent – and what better place to store all this information, than in your Car PC.

Vehicle Project Planner 2.0 has a Wizard driven interface that allows you to record new parts and new projects that take place on your vehicle.

You can then display the results of your project as graphs and reports.

Auto Organizer Deluxe

 http://www.primasoft.com/deluxeprg/auodx.htm

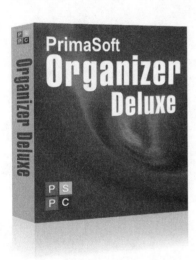

◀ Figure 8.6
Auto Organizer Deluxe box shot (courtesy Primasoft)

Primasoft Auto Organizer lets you combine all of your car related data into a single point of reference, allowing you to be in command of the information you need.

You can organize your automotive web resources: i.e. links to classic car clubs, car suppliers, my books on Amazon.com etc.

You can also organize your contacts – useful to have your address book while on the move, and even your expenses.

The software also allows you to put all of your trip information into a database so you can analyze trends.

Engine Performance Math Calculator

 http://www.virtualengine2000.com/

If you are an engine builder or modifier, you will find Engine Performance Math Calculator a great addition to your Car PC. The software allows you to quickly and simply calculate:

- Displacement
- Compression ratio
- Compression height
- Gearing, MPH
- Piston speed
- Horsepower
- RPM
- Intake CFM
- Injector sizing
- Loan payments
- Unit conversions
- ET and MPH

Best of all – the software is freeware and won't cost you a penny!

◀ Figure 8.7
Virtual Engine Calculator screenshot (courtesy Challenger Engine Software)

In addition to this, Virtual Engine 2000 also supply a range of other automotive calculates that are handy for the engine tuner and performance motor enthusiast.

- Dynamic Compression Ratio and Camshaft Selection Utility
- Virtual Dyno Standard
- Virtual Dyno Pro

Car Care

 http://www.carcaresoftware.com/desktop/top_intro.htm

In order to drive more efficiently you can get fuel economy statistics in both metric and imperial on your Car PC screen. You can then analyze the cost of your fuel over the same time period. By looking at trends in your fuel usage, you can learn how to drive more efficiently and – importantly – clearly see if your car needs some routine maintenance.

You can also import data from other sources with the software, so if you are migrating over from another application, all of your data can come with you.

You can look at how many miles your vehicles have done at a glance, or alternatively, look at how many hours they have been in use.

Furthermore, you can store that hard-to-recall information about your vehicle that you can never lay your hand on when you really need it, for example:

- Pain Code
- Key Code
- Horsepower Ratings
- Tire and Wheel Sizes
- Tire Pressures
- Loan Details
- Vehicle Identification Number
- Insurance Details
- Registration Information
- Car Stereo Keycode
- Auto Emission Statistics

Whether you are a shade tree mechanic, or car nut, you will find this software invaluable.

If you are into customizing cars, you can use the Car Care software to keep track of your expenditure, making sure projects stay on track and on budget.

Chapter 9

Getting more out of your Car PC

You've built your Car PC and got it to work, but wouldn't it be nice if you could do more with it? It's just the same feeling that you get after buying a new PC – you want to find what toys and extras you can add on to it. Thankfully, although you may be limited for space in a car, there are many accessories that you can get for your Car PC that take up minimal space and add a great amount of functionality to your Car PC. Let's take a look at some of your options.

Adding a webcam for reversing safely

You can fork out a fortune to have reversing ultrasonic sensors fitted to your bumpers, but at the end of the day you will always have blind spots, and there is nothing quite like seeing where you are going rather than relying on your "ears" to guide you.

Step in the webcam! When coupled to your Car PC, a webcam gives you eyes in the back of your head. Well, not quite, but it certainly allows you to see behind you without fitting a periscope in your Chevy.

The market is now saturated with a million webcams, all of which are very cheap and capable. I could recommend the Aiptek Pen Cam which is cheap and readily available through many online vendors. Its construction internally is very simple, meaning that you can hack it apart and mount it into your bumper or another suitable external enclosure.

Some of the Car PC front ends we have talked about will support webcams within their interface, allowing you to view the output of your webcam. However, if you want a stand-alone application that will allow you to view your webcam picture full-screen, you should check out Full Screen TV Viewer at

W [3] http://iridia.ulb.ac.be/~fvandenb/tools/tools.html

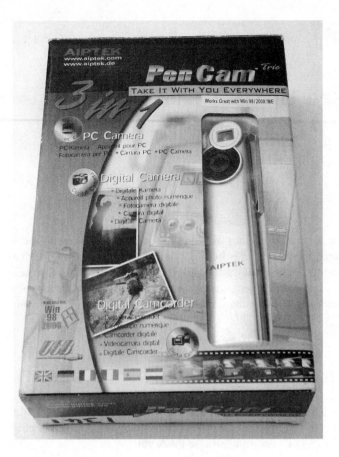

Figure 9.1
*The Aiptek Pen Cam
in box (Tim Watson,
Photomedia UK)*

Controlling your Car PC remotely with the Air Click remote control

Many car head units now come with a small fob-sized remote control. This can either be retained by the user or affixed to a steering wheel to provide fingertip control. Wouldn't it be nice if you could get some of these high-end features on your Car PC?

Thankfully … you can!

If it can be done with professional OEM equipment, there is no reason that it cannot be done with our homemade Car PC setup!

The Griffin Air Click remote control is a natty little unit with five buttons that allow you to take control of your Car PC. One of the advantages of the Air Click over many infra-red

Figure 9.2
*Remote control mounted on
steering wheel*

remote controls, is that the Air Click operates using wireless radio link enabling you to control your Car PC without having to point the thing at your Car PC.

Hardware installation is simply a matter of plugging the thing into any available USB port. Software is not hard to install.

Controlling your Car PC from your PDA with Total Remote

Another method of controlling your Car PC remotely, is by using your PDA. In the 1997 James Bond film *Tomorrow Never Dies*, Pierce Brosnan uses his Ericsson™ smart phone to drive his BMW.

In addition to this the phone had a fingerprint reader, key replicator and a stun gun. Why BMW never marketed this as a dealer fit option one can only wonder.

The key replicator and the stun gun might be a little out of most folks' ballpark, but the fingerprint recognition is certainly a feature on some high-end PDAs, and whilst driving your car with your PDA might not be desirable (or safe), controlling your in-car entertainment, navigation features and music certainly is.

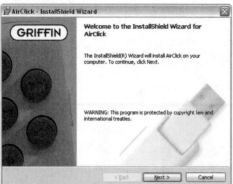

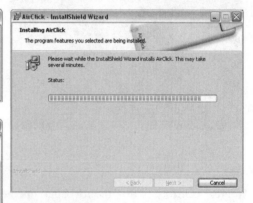

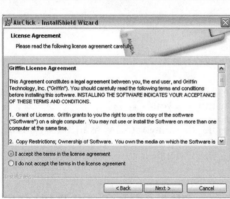

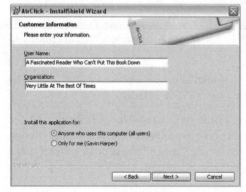

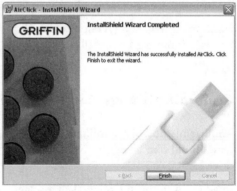

Figure 9.3(i-vii)
*Installation of Griffin Air Click
(courtesy Griffin Technologies)*

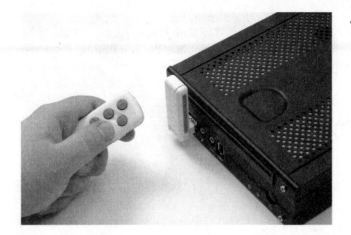

Figure 9.4
*Griffin Air Click USB
installed (Tim Watson,
Photomedia UK)*

PDAs are becoming a common sight in everyday life; young urbanites run their lives through them so there is no reason why you cannot run your car through them.

To accomplish this mean feat, you can use the infra-red port of your PDA with some clever software.

While this will certainly work, most folks' PDAs are fitted with relatively low power infra-red sending and receiving equipment as this method of data transfer is only really intended for close range.

There are two ways to solve this problem. One is pretty ugly! If you trawl the Internet, there are a number of people with sites that show how to retro fit high powered IrDa to the inside of your PDA. The main caveats of this method is that your warranty becomes as useful as a cat-flap in a submarine, and the chances are you will forget how your PDA went together in the first place, leaving you with a nice pile of techno-junk – an attractive paperweight maybe? And very little else.

There are also numerous disadvantages to using your Pocket PC's IrDa port – for a start it is entirely the wrong standard for remote control. It is possible for software to control the signal to send "remote control" information, but it is not incredibly effective because of the limitations of the system.

Enter the Griffin Total Remote (see Figure 9.5). The system from Griffin uses an entirely different approach. Your Pocket PC has a headphone socket to allow you to listen to music. Well, Griffin have hijacked this socket by producing an adaptor that plugs in with not one, but two infra-red LEDs (see Figure 9.6). In terms of power, this means that the Griffin Total Remote is likely to be twice as effective as your standard remote.

In tests with a Compaq Ipaq Pocket PC, Griffin have had this little baby do its business from over a hundred feet away. This means that your PDA is EASILY going to be able to control

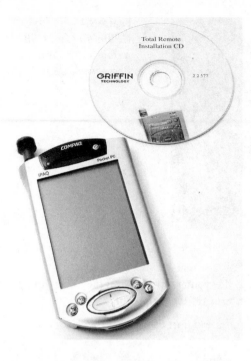

◀ Figure 9.5
The Total Remote package (PDA not included!) (Tim Watson, Photomedia UK)

◀ Figure 9.6
The Total Remote – up close and personal (Tim Watson, Photomedia UK)

devices while in the car, and the chances are, if you mount your Ir Receiver sensibly, you will be able to control your Car PC setup from outside your car.

The devices that are supported by the Griffin Total Remote are shown in the box below. If your PDA falls within this list then you are in luck. If your PDA is not included, then contact Griffin as it may have been added since this book went to print.

Compatible PDAs
Acer
Acer n20*
Acer n20w*

Asus
MyPal 620*

Audiovox
Maestro ARM*
Thera ARM**

Compaq
iPaq 31xx*
iPaq 35xx*
iPaq 36xx
iPaq 37xx*
iPaq 38xx*
iPaq 39xx*

Dell
Axim X5 (Range limited to less than 10 feet)
Axim X3 (Built in IrDA only)
Axim X3i (Built in IrDA only)

Fujitsu/Siemens
Pocket LOOX**

HP
Jornada 56x**
HP 19xx**
HP 22xx**
HP 41xx** (Built in IrDA only)
HP 43xx**
HP 51xx**
HP 54xx**
HP 55xx
HP 56xx**

NEC
MobilePRO P300*

Packard Bell
PocketGear 2060*
PocketGear 2030*

PC-ESolutions
MIA-PD600C*

Toshiba
Toshiba e31x
Toshiba e33x
Toshiba e35x*
Toshiba e4xx (Built in IrDA only)
Toshiba e5xx*
Toshiba e74x
Toshiba e75x (Built in IrDA only)
Toshiba e8xx (Built in IrDA only)
Toshiba Genio e550(GX/GS)*

UR There
@migo 600C*

Viewsonic
PDV35*
V37*

Zayo
A600*

Compatibility Notes:
*Not internally verified by Griffin. Should work based on hardware specs and existing system requirements.

**Compatibility reported by Griffin Technology customers.

**Due to hardware limitations this model will only work with the Griffin IR transmitter and not with the built in IrDA port. This means that learning and transmitting via built in IrDA will not work on this model. Also, it will require a 2.5 mm to 3.5 mm Stereo adaptor to accommodate our Griffin IR transmitter.

Total Remote installation

Installing Total Remote is no daunting task. After running the CD, click "Next" in the first window, and then agree to the license agreement in the second.

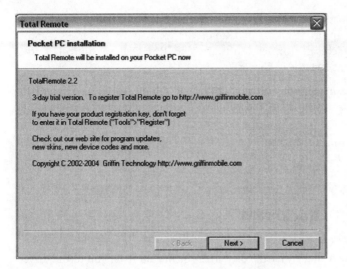

Figure 9.7(i)
*Figure Total Remote
Step 1 (courtesy Griffin
Technologies)*

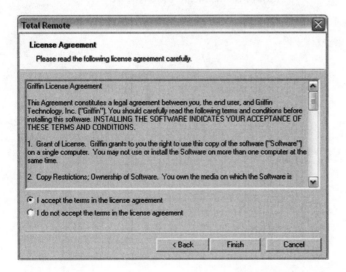

Figure 9.7(ii)
*Figure Total Remote
Step 2 (courtesy Griffin
Technologies)*

After this, your PDA's Active Sync software will take over and install the Total Remote software to your PDA. A soft reset to your PDA, and you are away.

Once you have installed the Total Remote, you need to add an infra-red receiver to your Car PC. There are a number of ways you can go about this – Microsoft are selling Remote Controls for their Media Center version of XP. Using Total Remote, you can map the key presses of the remote to your Total Remote software and control your Car PC using your PDA that way.

[3] http://www.microsoft.com/hardware/mouseandkeyboard/ProductDetails.aspx?pid=065

If you want to control a broader range of PC applications, then you need to nip down to www.evation.com to pick up an IRman. An IRman connects to your Car PC, and is supported by a number of pieces of front-end software. The IRman allows control of other pieces of software by mapping macros in Girder.

Another competitor is Streamzap, who offer a similar IR remote solution. Discard the remote and map the buttons to your PDA and you have another method of receiving Total Remote commands.

 http://www.streamzap.com/

Installing the Griffin RocketFM

Installation is a piece of cake: using either the driver CD, or the latest version of the drivers downloaded from http://www.griffintechnology.com/ install the software.

First of all, you will get the InstallShield Wizard start screen, with a cutesy watermarked picture of RocketFM in the background.

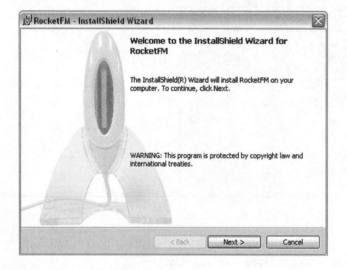

Figure 9.8(i)
Griffin RocketFM software installation Step 1 (courtesy Griffin Technologies)

Click "Next" to move onto the next screen.

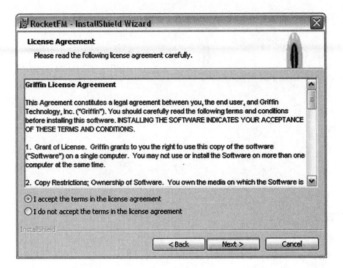

Figure 9.8(ii)
*Griffin RocketFM
software installation
Step 2 (courtesy Griffin
Technologies)*

You are now going to need to read through the license agreement and click "I accept" in order to move on to the next step.

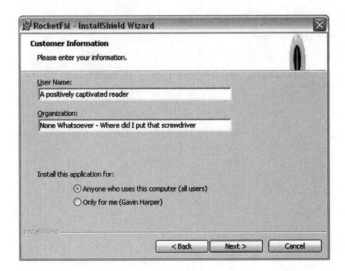

Figure 9.8(iii)
*Griffin RocketFM
software installation
Step 3 (courtesy Griffin
Technologies)*

Next enter your name and the organization you work for or belong to. If the Car PC is for personal use, you might like to put "Organization" as a Car Club that you belong to.

Figure 9.8(iv)
*Griffin RocketFM
software installation
Step 4 (courtesy Griffin
Technologies)*

Now the installation will take care of itself for a moment. Wait a while.

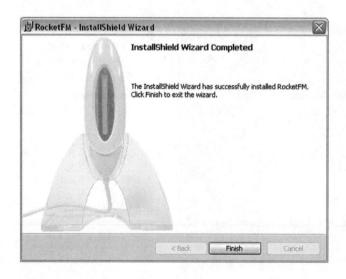

Figure 9.8(v)
*Griffin RocketFM
software installation
Step 5 (courtesy Griffin
Technologies)*

And now click "Finish."

The software is now on your Car PC.

Hardware installation is also very painless – simply plug the device into a spare USB port.
See Figure 9.9.

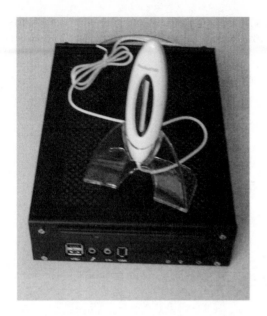

Figure 9.9
The RocketFM installed

Hint

RocketFM works much more effectively when the volume on your Car PC is turned up to maximum. You can then use the volume control on your car stereo to adjust the volume.

You should now tune your car stereo so that it is operating at the RocketFM's default frequency of 88.1 FM.

Figure 9.10
Car stereo tuned to 88.1 FM

If you are not receiving a signal from your Car PC you need to configure your Car PC to make sure that the audio is being output with "RocketFM" as the output device. To do this you need to go to:

Start > Control Panel > Sounds and Audio Devices

…and click the "Audio" tab… This will bring up the window as seen in Figure 9.11.

Figure 9.11
Sounds and audio properties window (courtesy Griffin Technologies)

Under "Sound playback," you will need to set the default device as "RocketFM."

If you want to change the frequency that RocketFM broadcasts on, go to the RocketFM Control Screen.

This will bring up the RocketFM Control Screen (see Figure 9.12). The control screen allows you to set whether the RocketFM will transmit in stereo or mono, and what frequency it broadcasts at. You can use the full spectrum of FM frequencies from 88.1 FM to 107.9. The unit will broadcast up to 30 feet.

Figure 9.12
*The Griffin RocketFM control screen
(courtesy Griffin Technologies)*

Great Idea

This is great news if you want to use your vehicle at a car show as a mobile jukebox – simply queue up all your music files as a play list, and broadcast to nearby P.A. equipment.

> **Tip**
>
> You may find that the default frequency of 88.1 FM is already allocated to a commercial radio station in your area. In this case, you will need to tune your RocketFM to use another unallocated frequency. You can find a spare frequency by using the "Manual Tuning" option on your car stereo to find a station that just sounds fuzzy with no talking or music. Note the frequency, and set your RocketFM to use it.

Troubleshooting your RocketFM

Q. When I am listening to music, I hear distortion on my car stereo. Is there any way to get rid of this.

A. You may need to turn the "Master Volume" on your Car PC down a few notches. While the RocketFM performs better when the volume is set to max, you may find it is being "overdriven" by the Car PC, causing distortion. By turning the volume down on your Car PC, you can compensate by turning the volume up on your car stereo to the limits of your system (and tolerance of your neighbors).

Q. I can hear audio through my car stereo, but not through the "line out port."

A. Your Car PC can only drive one audio output at a time.

Q. I can't hear Car PC audio on my car stereo.

A. First of all check the tuning; just a single digit out of place can make a lot of difference. You might find that you need to change the location of your RocketFM – there is a lot of metal in a car, and metal acts as a "shield" for radio frequency. Try moving your RocketFM so it is nearer to your car aerial. Next thing to do is unplug and plug in your RocketFM. Windows may not have recognized it the first time. Finally, if you are using your RocketFM through a USB hub, try plugging it directly into your Car PC.

Replacing your radio

For many folks finding space for a Car PC may mean sacrificing the radio, but at the end of the day this is no bad thing, and it doesn't mean that you have to sacrifice functionality, the reason being – all of this can be replaced by adding some suitable hardware to your Car PC.

Enter the Griffin Radio Shark, an AM/FM radio fully tunable through supplied software. And the even better news is that buying something else from Griffin means *more* coolable LEDs. Many more and your vehicle will look like a police car on a call.

Griffin Radio Shark

Installing the Griffin Radio Shark is simplicity itself. It has the advantage over many AM/FM tuners on the market of not requiring a separate aerial and so provides an easy, simple "out of the box" solution.

Installing the software is simplicity itself, the Radio Shark comes with an installer – after the software is installed you can simply plug in the hardware and get on with things.

To install the Radio Shark hardware, plug the device into one of your Car PC's USB ports. Front or rear, it makes no difference.

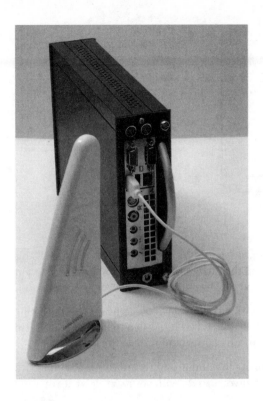

◀ Figure 9.13
*Griffin Radio Shark connected to Car
PC (Tim Watson, Photomedia UK)*

WINDOWS SHORTCUTS

Space: Mute
F9: Pause (affects timeshift as well as playback)
ESC: Close
Up Arrow: Volume Up
Dn Arrow: Volume Down
Left Arrow: Previous Preset
Right Arrow: Next Preset
Ctrl-Left Arrow: Search Left for Station
Ctrl-Right Arrow: Search Right for Station

When you first fire up the Radio Shark, you are going to be presented with the following screen (see Figure 9.14).

Figure 9.14
The Griffin Radio Shark "Home Screen" (courtesy Griffin Technologies)

This screen has all of the basic controls you need in an FM radio. My only criticism of this piece of software is that it can come out a bit on the titchy side if your screen is set to a decent resolution. However, if you use a stylus you will find even the small controls a breeze.

Next, picture this. You are about to go into work but there's a great game on and it is being broadcast by radio. You know that you can sneak out at lunch to listen to it, but the problem is the game is at 11 a.m. Should you leave that early you would raise suspicion!

No worries. With the Radio Shark you can pause live audio, and record shows to come back and listen to them later! Yet something else that you can do with your Car PC that you can't do with a conventional car stereo!

Programming your Car PC to record radio is thankfully *much* easier than setting up the VCR. Go to the scheduling window (Figure 9.15)

Figure 9.15
Griffin Radio Shark scheduling window (courtesy Griffin Technologies)

Here you can set up what programs you want to listen to. Just click on "add" and you bring up the event window. Thankfully, all the details here are pretty much self-explanatory. Enter the time, station and audio format and you are away! Now bring up the event window, Figure 9.16.

Figure 9.16
*Griffin Radio Shark event
window (courtesy Griffin
Technologies)*

Add a twiddly knob to your Car PC with the Griffin Powermate

Your Car PC's front does look rather fetching with its sleek black front and slot loading drives. However, it does rather lack an array of buttons to press and fiddle with. This may come as a dismay to many readers who relish the prospect of telling their kids off for playing with the radio on a long journey. Furthermore, the anxious among you might like something to fiddle and play with during those nervous moments.

Thankfully, there are even *more* ways of controlling your Car PC's capabilities!

BMW spent a shed load of deutschemarks (or would that be euros?) on researching how they could make a car's controls more intuitive, and in true research fashion, after compiling many figures came up with the type of observation that any competent 7-year-old could make – people like rotary knobs.

Roll on the iDrive

Whilst the Griffin Powermate may not have the functionality of a BMW iDrive, it certainly adds a rotary knob to your Car PC, and if you mount it somewhere prominent adds a touch of cool to your car with its space age blue LEDs.

Most people are used to turning up their car stereo's volume by simply turning a knob, without having to navigate any complex menus or clicking on pointers. To give your Car PC that intuitive feel, why not install a Griffin Powermate? (see Figure 9.17.)

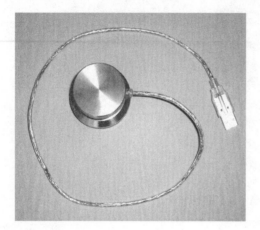

Figure 9.17
Griffin Powermate Rotary Controller Hardware

What is a Powermate, you might ask? A Powermate is a multifunctional rotary controller. Any wiser? No? Then I'll explain.

The Griffin Powermate is a smart aluminum knob surrounded by a skirt of gorgeous blue LED light. This thing will look cool in any car interior.

You can turn the knob clockwise and anticlockwise, as well as being able to push it in.

Great! You say – but what can I do with it?

The Powermate connects to your Car PC using a standard USB interface. Software that you install allows you to map the movements of the Powermate to variables within programs. For example, you might like to use Powermate in Windows Media Player to turn the volume up and down when you turn the knob, and mute it when you press it in. In a Car PC setting, this means that you can be listening to your tunes, and then quickly mute the Car PC when you receive an incoming phone call on your cell phone. Another useful application for the Powermate in a Car PC setting is to be able to Zoom In and Zoom Out in a GPS application, allowing you to both see the detail of the street you are on, and your position relative to your destination with relative ease.

And the best thing is – installation is a breeze.

Installing the Griffin Powermate

Start the software on the CD or download the latest version from

 www.griffintechnology.com

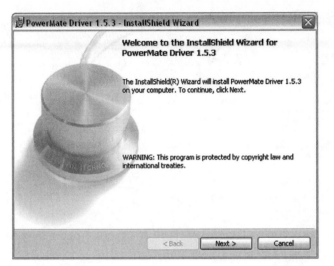

Figure 9.18(i)
Griffin Powermate
software installation
Step 1 (courtesy Griffin
Technologies)

The first screen you will be presented with is this one, simply click "Next" to move on to the next screen.

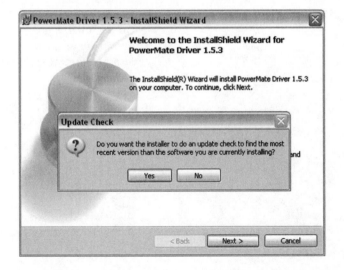

Figure 9.18(ii)
Griffin Powermate
software installation
Step 2 (courtesy Griffin
Technologies)

You will then be prompted to check for an update to see if there is any more recent software. To check for an update, you will need an Internet connection. If you have an Internet connection, it is best to click "Yes"; if not, then click "No" as this step is irrelevant.

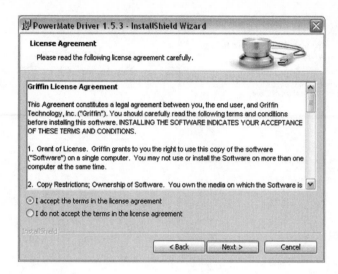

Figure 9.18(iii)
*Griffin Powermate
software installation
Step 3 (courtesy Griffin
Technologies)*

You now need to agree to the licensing and click "Next." Read the agreement and then select the "I accept" button.

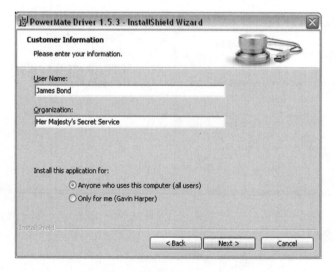

Figure 9.18(iv)
*Griffin Powermate
software installation
Step 4 (courtesy Griffin
Technologies)*

Enter your name and organization in the spaces provided and click "Next."

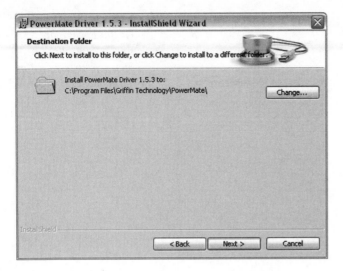

Figure 9.18(v)
Griffin Powermate
software installation
Step 5 (courtesy Griffin
Technologies)

If you are happy with the folder selected (and there is no reason why not) select "Next."

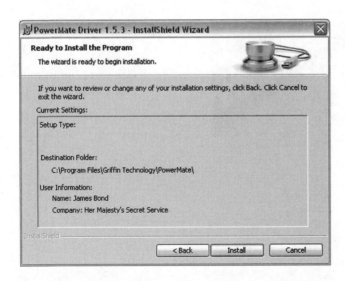

Figure 9.18(vi)
Griffin Powermate
software installation
Step 6 (courtesy Griffin
Technologies)

Confirm the details displayed on the screen and click "Next."

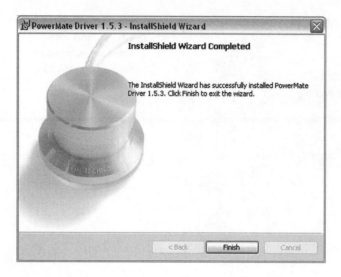

Figure 9.18(vii)
Griffin Powermate
software installation
Step 7 (courtesy Griffin
Technologies)

That's it! You are all done now.

Plug your Griffin Powermate into a vacant USB port. The PC should recognize it because of Plug and Play and install it automatically (Figure 9.19).

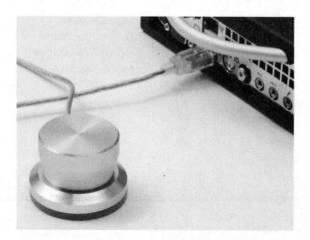

Figure 9.19
The Griffin Powermate
plugged into a spare
USB port (Tim Watson,
Photomedia UK)

Hint

You can connect your Powermate to any USB port, whether this is through a hub or other device. However, you can only use the Powermate as a "Power On" knob when it is connected directly to your Car PC.

When you are finished, you should see a new icon in the taskbar. Click on it to bring up the "Powermate Properties" Window (Figure 9.20).

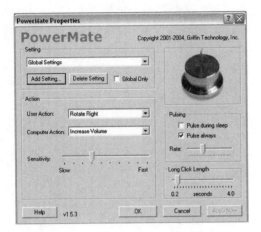

Figure 9.20
Powermate Properties (courtesy Griffin Technologies)

Chapter 10

In-Car GPS

GPS: the background information and history

On October 4th 1957, the Soviet Union launched their satellite Sputnik "Спутник."

Meaning "traveling friend," the satellite started the space race, a series of events that would eventually enable us to navigate using satellites. There is a nice analogy here … as when you are driving along the road on your own with the sat-nav being the only thing in the vehicle talking to you, that piece of equipment can seem like a "traveling friend."

Anyways, back in the days of Sputnik, two researchers, Guier and Wafenback, worked out a way to track the position of the satellite by looking at the Doppler shift in the frequency transmitted by the satellite. If you don't know what Doppler shift means think of it this way: an ambulance approaches from the distance, its siren blaring "nee naw nee naw." As the ambulance approaches, the frequency is high; however, as the ambulance passes you notice a drop in frequency. What is happening here, is that as the ambulance approaches you, the sound waves are in effect being compressed as the "peaks" of the sound wave reach you quicker because of the movement of the ambulance. As the ambulance drives away from you, the waves are being "rarified" because the "peaks" are now more spread out due to the vehicle's movement causing a drop in frequency. This approach, analyzing the change in frequency, was used to track Sputnik.

Some years later, a bright spark called Frank McClure realized that if you invert the Sputnik scenario, i.e. fix the satellite in a permanent position, then you can determine your position by applying the reverse logic to determining Sputnik's position. This was used to allow nuclear subs to track their position.

As ever, if the government are going to spend shed loads of money, you know they are going to spend it on things designed to blow people up rather than save lives.

In 1964, a system of five satellites was launched called "TRANSIT." This was the first series of dedicated "Global Positioning" satellites that were launched.

Later – in 1967 and 1969 – the "TIMATION" satellites were launched that used a more advanced approach; quartz clocks on board the satellites would relay precise time information to earth.

Between 1968 and 1971, 621B was launched. This was a development upon its predecessors in that an element of random information was introduced to the system rendering it useless for accurate information to anyone other than the military who held the codes to this random information. (A nice touch to a system that was paid for by tax payers don't you think?)

On December 17, 1973 the NAVSTAR system was approved. This was a conglomeration of all the "best" bits of the above system (that includes the random codes which rendered the system useless to you or me unless you could tolerate a tolerance of 50 meters) – a nice way to thank the US public for all the dollars they spent on it.

Now there are 25 GPS receivers in orbit, and thanks to Bill Clinton, who on the 1st of May, 2000 said:

> The decision to discontinue Selective Availability is the latest measure in an ongoing effort to make GPS more responsive to civil and commercial users worldwide... This increase in accuracy will allow new GPS applications to emerge and continue to enhance the lives of people around the world.

Finally the US public could get value for money out of the system they paid for in the first place. GPS information is freely available; there are no licensing fees to pay and the technology is in the public domain.

The working group that developed the NAVSTAR system set out with a goal of reducing the cost of the receiver technology to $10,000 a piece. Now, the shrewd shopper can pick up a second-hand GPS mouse on eBay for $10.

How does the technology work?

GPS uses a technique called trilateration. Each GPS/Navstar satellite has onboard a highly accurate atomic clock which transmits a time signal as well as an identifier. On earth our GPS receiver takes the received signals of all the satellites that are "in view" and calculates the discrepancy between the different time signals. By analyzing the delay between the transmitted time and the time received (done by comparing against an accurate internal clock which is adjusted by averaged time signals from all the satellites) it is possible to work out how far away each satellite is. By drawing an imaginary "sphere" around the satellites, we can calculate mathematically where the "spheres" intersect of all the received signals to work out our position. The earth forms another point of reference, as it is spherical and allowing us to work out all the possible points and what ones are "sensible" values. This is illustrated in Figure 10.1.

This system is augmented by WAAS, which is a means of increasing the accuracy of the GPS information using ground based radio stations.

Links to sites of interest with GPS software

Home of CoPilot Software

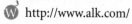 http://www.alk.com/

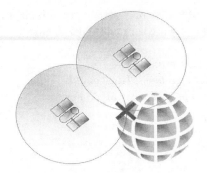

Figure 10.1
How GPS works diagram

Home of GPSS Free GPS Software

W³ http://www.gpss.co.uk/

Home of MS Autoroute

W³ www.microsoft.com/uk/homepc/autoroute

Home of MS Mappoint

W³ www.microsoft.com/mappoint

Home of MS Streets and Trips

W³ www.microsoft.com/streets/

Home of Destinator GPS Software

W³ www.destinator1.com

Home of "Map Monkey" a touchscreen compatible front end for Destinator

W³ www.mapmonkey.net

Choice of GPS hardware

If you are happy to sacrifice your COM port, you should be able to pick up a very good deal on a serial GPS receiver. This will not require any drivers and as long as the protocol is supported by your GPS software you will be all right.

USB GPS receivers are also very cheap. Devices based upon the BU-303 chipset, like the one sold by the mp3car.com store in Figure 10.2 have become ubiquitous and are very reliable.

These incorporate a piece of driver software which creates a "virtual" serial port, and maps the information from the USB interface to this port for easy use with most GPS software.

◀ Figure 10.2
A USB mouse based on the BU-303 chipset (courtesy Mp3car.com)

Also very common are the newer range of Bluetooth GPS devices which require no physical connection to provide GPS information. This is good, as it means that you can always position the device for optimum signal without having to worry about routing cables.

If you shop around, you may even find some GPS software such as Microsoft Streets & Trips 2005, which is available with a GPS mouse (Figure 10.3). Although these bundles do not *always* represent the best deal on the market, you can guarantee an easy setup. Saying that, most GPS mice on the market are NMEA compliant, so you should have little trouble whatever mouse you buy.

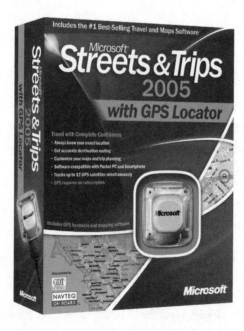

◀ Figure 10.3
Microsoft Streets & Trips (courtesy Microsoft)

Some links you might want to look at for GPS hardware

- www.garmin.com
- www.deluo.com
- www.delorme.com

Choice of GPS software

One of the factors that might influence your choice of GPS software is how it integrates into your front end software. Microsoft Map Point (see Figure 10.4) has been popular with many folks developing front ends because it is very easy to work with using Microsoft Visual Basic.

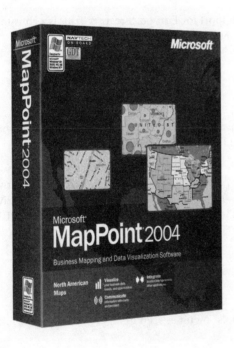

Figure 10.4
Microsoft Map Point (courtesy Microsoft)

Other front end developers have chosen to work with Destinator which has proven especially popular in European countries due to its good mapping support. Destinator also comes with some really great 3D visualizations of the road ahead.

There is a wide range of GPS software out on the market. One of your main concerns when choosing an item of GPS software is what locality it was designed for. U.S. users report good success with Copilot products. Figure 10.5 shows Copilot 9.

Figure 10.5
Copilot Live (courtesy Mp3car.com)

Destinator, as mentioned earlier, has good support for Europe. Streets & Trips is a consumer oriented US product, whilst Autoroute is its European equivalent.

Typical GPS installation

If you have an NMEA compliant GPS device that uses your serial port, you will not need to worry about installing drivers; your installation should communicate with the serial GPS directly.

Installation will vary depending on what GPS receiver you have bought. If you have one with the BU-303 chipset, the driver you install is not a "GPS" driver as such, but in fact a driver for the chip which converts the serial signals from the chipset to a USB interface. The driver you install on your Car PC creates a "virtual" COM port and maps the signal from the USB interface to this virtual COM port.

Once you have the drivers (if any) installed, you will need to tell the GPS software what COM port you are using (whether it be real or virtual).

I am going to talk you through as if you were using Autoroute as your GPS software.

First open up the software and go to the "Tools" menu as seen in Figure 10.6.

Figure 10.6
*Autoroute GPS setup Step 1
(courtesy Microsoft)*

From the "Tools" menu, you need to select the "GPS" option, and from the window this brings up select "Configure GPS receiver."

▼ Figure 10.7
Autoroute GPS setup step 2 (courtesy Microsoft)

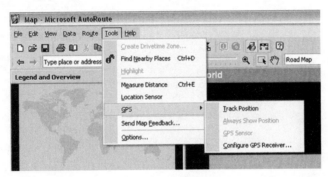

This brings up the GPS receiver setting windows shown in Figure 10.8. Here you need to select what COM port your GPS device is connected to from a list of the available ports.

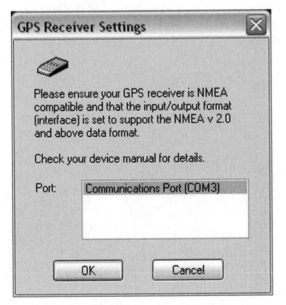

◀ Figure 10.8
Autoroute setup Step 3 (courtesy Microsoft)

Epilogue

The Car PC market can only take the lead in the Home PC market. Already we are seeing dual core processors and 64 bit data buses in mainstream PCs, and it is only a matter of time before the technology becomes scaled down in size and suitable reductions are made in power consumption for it to be economical to squeeze the bags of extra power into a Car PC.

New operating systems are released by the year, and software continually evolves to take advantage of the increased capabilities of the hardware.

Additionally, the issue of computers in cars has increasing relevance in these days of energy shortages and environmental concerns. Already, embedded computers have made great strides in reducing emissions and increasing fuel economy. How long will it be before we see more and more of the "hard wired" functions of the car transferred to software, which requires minimal power and minimal materials?

By using sophisticated PC intelligence to plan our journeys, we can help save the environment by lowering fuel consumption. By taking more direct routes and avoiding traffic, we can ensure that we get the best out of our vehicles and traveling experience.

Navigation can only get better in the years to come. Already the U.S. global positioning system works well in most terrains; however, with systems from Europe, China and Russia coming on-line in the next decade, we can expect to see integrated receivers that pick up all signals and work out the best position fix by assimilating all the data.

One thing is certain, the Car PC hobby is quickly gathering pace; already we are seeing a lot of niche products designed to cater to this market, and it is only a short while before these products become mainstream and there are a wide variety of Car PC products in your average computer store.

Start now, and you can say "I had a Car PC before they were mainstream!"

Appendixes

Appendix A

Suppliers' Index

Andrea Electronics Corporation

65 Orville Drive
Suite One
Bohemia, NY 11716
Tel: 0800-442-7787
 http://www.andreaelectronics.com

Blue Sunset Software
 http://www.milemate.com

Challenger Engine Software, LLC

115 Jeanette Drive
Granite City, IL 62040
Sales@virtualengine2000.com
Techhelp@virtualengine2000.com

Crucial Technology

3475 E. Commercial Ct.
Meridian, ID 83642
Toll-free for USA & Canada: 800-336-8915
Tel: 208-363-5790
Freephone: 0800-013-0330
International: +44-0-1355-586-100
 www.crucial.com

Digital Worldwide

customerservice@digitalww.com
sales@digitalww.com
Tel: 847-546-5822
 www.digitalww.com

Griffin Technologies

Griffin Technology
1930 Air Lane Drive
Nashville, TN 37210
Tel: 615-399-7000
Fax: 615-367-6468
 http://www.griffintechnology.com/contact/email.html

Lone Wolf Software

sales@lonewolf-software.com
support@lonewolf-software.com
 www.lonewolf.software.com

Microsoft Corporation

One Microsoft Way
Redmond, WA 98052-7329
Tel: 425-882-8080
Fax: 425-706-7329
 Web site: http://www.microsoft.com

Mp3car.com Store

 www.mp3car.com/store

Özen Elektronik

Ali Nihat Tarlan Cad.
2. Bahceler yolu 13/7
TR 34744 Bostanci, Istanbul
Turkey
Tel: 0090-216-445-5882
Fax: 0090-216-445-5883
info@ozenelektronik.com
sales@ozenelektronik.com
support@ozenelektronik.com
 www.ozenelektronik.com

Primasoft

PrimaSoft PC, Inc.
P.O. Box 456
Surrey, BC V3T 5B7
CANADA
Tel: 604-951-1085

support@primasoft.com

 www.primasoft.com

Travla

2F., No.16, Lane 77, Sing-ai Rd., Neihu District,
Taipei City 114,
Taiwan
Tel: 886-2-2793-8205
Fax: 886-2-8791-7285
4755 Hannover Place
Fremont, CA 94538
Tel: 510-656-5475 ext: 114
Fax: 510-656-5477
sales@Travla.com

 www.travla.com

Appendix B

Car PC Forums and Clubs

Car PC Info

W³ http://www.car-pc.info
German forums for Car PC.

Car TFT Forum

W³ http://www.cartft.com/community/carpcforum/index_html
A smaller forum with a couple of thousand posts on all the regular subjects.

Digital Car Forums

W³ http://www.digital-car.co.uk
Good forums targeted at UK/EU (European Union) users.

Mp3Car.com

W³ www.mp3car.com
Still the original and best site for In-Car Computing. International appeal.

VIA Arena Car PC Forum

W³ www.viaarena.com
Website of the motherboard manufacturer VIA. A small but growing forum on Car PC issues, and many other discussions about Mini-ITX issues. Author writes regularly for the News section of this site.

Yahoo Groups

Car Computing

Ⓦ http://groups.yahoo.com/group/carcomputing
Small Yahoo group dedicated to car computing.

Car PC

Ⓦ http://autos.groups.yahoo.com/group/carpc
A small Yahoo group with few posts.

OBD2

Ⓦ http://autos.groups.yahoo.com/group/obd2
Yahoo group dedicated to the OBD-2 interface standard, and how to use it with a computer to solve diagnostic issues.

South African Car PC

Ⓦ http://groups.yahoo.com/group/sacarpc
A group for South African Car PC enthusiasts to get together.

Appendix C

Car PC Installs

Some good websites to check out enthusiasts' Car PC Installs.

Audi Rueger
Ⓦ http://audi.rueger-net.de
A German site showing installation of a Car PC in an Audi.

Autohacks
Ⓦ http://autohacks.net/news/2006/01/16/nissan-350z-car-pc-install
Nissan 350Z Car install.

Carputer Club
Ⓦ http://www.carputerclub.com
Car PC integrated into a Sony 10-disk changer box.

Carpc Maniyax
Ⓦ http://carpc.maniyax.jp
Japanese Car PC site.

Carpc Nl
Ⓦ http://www.carpc.nl/index.shtml
Nice Dutch Car PC website.

Dakota Project
Ⓦ http://www.dakotaproject.com
A basic Car PC install in a Dodge Dakota.

Fdzmm

W³ http://www.fdzmm.nl/ducarputer

This site shows installation of a Car PC in a Fiat Ducato – why spoil a nice Car PC?

Lets Communicate

W³ http://www.letscommunicate.co.uk/carpc

Incredibly smart site featuring an install in a Vauxhall Vectra.

Members Iinet

W³ http://members.iinet.net.au/~cpkh/carpc

Australian site showing installation in a Subaru.

Mp3 Carisma

W³ http://www.mp3carisma.com

Install in a 1997 Mitsubishi Carisma.

Neoslogic

W³ http://www.neoslogic.com/media4d

Professional looking Car PC install in a single DIN bay.

Project Astra

W³ http://www.projectastra.co.uk

Smart site showing installation in a Vauxhall Astra.

Staf Org

W³ http://www.staf.org.uk/car-puter

Site dedicated to building a "budget" Car Computer.

Stevie G

W³ http://www.stevieg.org/carpc

Nicely integrated clean install in a 1993 Ford Focus.

That Strife

(W) http://www.thatstrife.com/carproject

The new home of "This Strife," good Car PC project.

Time Killer

(W) https://www.timekiller.org/carpc/index.php

A really nice Linux-based install with good integration into the car's cockpit.

Webrings

(W) http://g.webring.com/hub?ring=mp3car

The mp3 Car Webring.

(W) http://j.webring.com/hub?ring=thenewmp3carwebr

The new mp3 Car Webring.

(W) http://n.webring.com/hub?ring=automotivecomput

Automotive Computers Webring.

Appendix D

Car PC Software Links

Here are some links to popular Car PC front end software. The software featured is developed by members of the community rather than commercial organizations so the quality can be variable. Also some of these projects have not been updated for some time, but then again, many users still find them very functional.

Autotouch
(W)³ http://www.sarinarts.com/autotouch.htm
Older Car PC interface with no mp3 support.

Centrafuse
(W)³ www.centrafuse.com
Well-developed Car PC front end with support for plug-ins.

Frodoplayer
(W)³ www.frodoplayer.com
Well-established software. Frodobaggins has a good reputation on the mp3car.com forums.

Media Cruiser
(W)³ www.media-cruiser.com
Old Car PC software.

Media Engine
(W)³ www.mediaengine.org
The original and best.

Mobile Impact

W³ http://thequiltcupboard.net/mobileimpact/download.htm
Sleek media software with split screen support.

Mobile Media Center

W³ http://www.hybrid-mobile.com
Does exactly what it says on the tin!

Mobilus

W³ http://sourceforge.net/projects/mobiluspc
Work in progress Car PC shell written in C#.

NeoCar Media Center

W³ www.neocarmediacenter.com
French Site for NeoCar Media Center.

PyCar

W³ pymedia.org/pycar/
Car Media Center written in Python.

Roadrunner

W³ http://guino.home.insightbb.com/roadrunner.html
Based on the old Medicar.

Silverwolf AES

W³ http://silverwolf.intuitionsys.com
Car PC software designed with Gentoo Linux in mind.

Index

About the Author

Gavin Harper is a member of a number of car clubs and online automobile hobbyist/enthusiast groups. He writes regularly for VIAarena.com, and started their Car PC Forum. Author of *50 Awesome Auto Projects for the Evil Genius* (also from McGraw-Hill), as well as other forthcoming titles in the Evil Genius series, he has been building car computers and other automotive projects for years. He lives in Essex, United Kingdom.

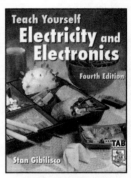